T0177794

Open Ecosystems

Open Ecosystems

ecology and evolution beyond the forest edge

William J. Bond

OXFORD
UNIVERSITY PRESS

OXFORD
UNIVERSITY PRESS

Great Clarendon Street, Oxford, OX2 6DP,
United Kingdom

Oxford University Press is a department of the University of Oxford.
It furthers the University's objective of excellence in research, scholarship,
and education by publishing worldwide. Oxford is a registered trade mark of
Oxford University Press in the UK and in certain other countries

© William J. Bond 2019

The moral rights of the author have been asserted

First Edition published in 2019
Impression: 1

All rights reserved. No part of this publication may be reproduced, stored in
a retrieval system, or transmitted, in any form or by any means, without the
prior permission in writing of Oxford University Press, or as expressly permitted
by law, by licence or under terms agreed with the appropriate reprographics
rights organization. Enquiries concerning reproduction outside the scope of the
above should be sent to the Rights Department, Oxford University Press, at the
address above

You must not circulate this work in any other form
and you must impose this same condition on any acquirer

Published in the United States of America by Oxford University Press
198 Madison Avenue, New York, NY 10016, United States of America

British Library Cataloguing in Publication Data
Data available

Library of Congress Control Number: 2019944900

ISBN 978–0–19–881245–6

DOI: 10.1093/oso/9780198812456.001.0001

Printed in Great Britain by
Bell & Bain Ltd., Glasgow

Links to third party websites are provided by Oxford in good faith and
for information only. Oxford disclaims any responsibility for the materials
contained in any third party website referenced in this work.

Preface

Open ecosystems are an anomaly, an aberration from the way nature is supposed to work. They should be forests, the natural successional climax where the climate is warm enough and wet enough for tall plants to grow and shade out the smaller ones. For at least 200 years, open ecosystems have been explained away, by science, as degraded, deforested landscapes. Only in the last 20 years have we begun to recognize our error and accept that many of these ecosystems are ancient, far too old to be created by human actions. With that recognition, we have also had to accept that the enemies of forest, wildfires and large herbivores, are also ancient and powerful allies of the biota of open ecosystems. This book introduces readers to the major topics that are being explored in research on open ecosystems. From small, recent beginnings, an avalanche of research papers are being published covering diverse aspects of the ecology and evolution of open ecosystems. I have been involved in several of the topics and have tried to capture their brief history and where they are heading. This is no easy task given the rate at which new work is appearing.

There is an enormous literature, including a number of books, on each of grasslands, savannas, Mediterranean-type shrublands, and a rapidly exploding literature on fire as a major regional and global force. But this is the first book, I think, that takes a global view on all of these ecosystems and explores common attributes and explanations for their existence. My constant difficulty in writing it was trying to avoid getting side-tracked into the rich and familiar literature on fire ecology, or on the ecology of the major open biomes, grasslands, savannas, or shrublands. I have tried to stick to the fundamental questions of where, why, and for how long open, typically low biomass ecosystems occur side by side with closed forest ecosystems. In trying to look at the world afresh, perhaps the most unnerving discovery for me was the profound way in which organisms can alter their own environments to better suit themselves and their progeny. I had been exposed so long to the view that the physical climate sets the rules to which plants and animals ultimately have to conform. It is challenging, to say the least, to accept that completely different ecosystems can develop from the same physical environment, and that life can manipulate the raw materials in such profound ways.

I apologize for glaring omissions of what those in the know consider key papers. Nevertheless, I hope there are enough signposts to get the uninitiated started if they want to purse any research area further. As regards geographical coverage, I have tried to take a global perspective since I believe that recognition of open ecosystems around the world will force a major revision of our understanding of biogeography at a time when we most need it. I am most familiar with southern African grasslands, savannas, and shrublands on the light side and with closed evergreen forests and thickets on the dark side. I have been fortunate in visiting, and often working in, open ecosystems in the southern, and parts of the northern, hemisphere where I have tried out the ideas developed in this book. It has been an exciting and enjoyable endeavour shared with many excellent hosts.

A major motivation for writing this book was my concern over internationally supported policies and funding for eliminating open ecosystems by 'reforesting' them. There are areas of the world in dire need of protection of existing forest and restoration of felled forests. But it is alarming to see proposals taken seriously for planting up trees in an area half the size of the Australian continent in the next decade, supposedly to reduce global warming. Fanatical afforestation on this scale could cause untold damage for generations to come. At the very least, people must learn to tell the difference between ancient old-growth open ecosystems and true deforestation so they know what they are destroying. It is my hope that this book will contribute to wiser counsel.

I would like, first, to thank Ian Woodward who was a pioneer in devising tests of the global importance of climate in shaping vegetation. Guy Midgley and I spent many hours stranded in an airport by a snowstorm, putting together ideas which Ian helped us develop. Ian also educated us on global geography, plant physiology, simulation modelling, and fine culinary experiences. Bill Robertson and the Mellon foundation played a huge part in supporting research and student training while encouraging journeys beyond the usual boundaries. Jeremy Midgley has been a research associate for most of my academic career, an inspiring colleague full of new ideas balanced by scepticism for the accepted order of things. Jeremy, Caroline Lehmann, and Will Stock read parts of the book and provided valuable feedback. I have benefitted greatly from colleagues at UCT including Mike Cramer, Will Stock, Ed February, Timm Hoffmann, Lindsey Gillson, Muthama Muasya, and, beyond UCT, Michelle van der Bank, Jonathan Davies, and Brad Ripley. I have had the privilege of working with excellent students, postgraduates, and post-docs, many of whom are now research colleagues. Particular thanks to Sally Archibald, Carla Staver, Julia Wakeling, Gareth Hempson, Corli Coetsee, Ben Wigley, Barney Kgope, Nicola Stevens, Vhali Khavhagali, Heath Beckett, and more. Tristan Charles-Dominique spent a very productive post-doc with me during which he made great progress in exploring functional traits in the context of green, black, and brown worlds. I am indebted to him for his insights and artwork. My colleagues during my time with SAEON broadened my horizons while giving me room to develop my enthusiasms. Special thanks to Johan Pauw, Beate Holscher, and Nicky Allsopp.

Part of the excitement of the science of open ecosystems is that it is a global endeavour with new ideas and observations changing the way we look at nature in many parts of the world. It can be very difficult picturing vegetation from the literature—e.g. there are many shades of meaning to the word 'forest'. Caroline Lehmann hosted a breakthrough field trip and workshop in Australia that brought together colleagues from South America, Africa, and Australia during which we developed a common vocabulary and set of concepts. On my travels, I am very grateful to the following for generous hospitality: Dick Williams, Garry Cook, Alan Andersen, Dave Bowman, and Grant Wardell-Johnson in Australia; Alan Mark, Kath Dickinson, and Bill Lee in New Zealand; Bill Hoffmann, Giselda Durigan, Alessandra Fidelis, Inara Leal, Geraldo Fernandes, and Elise Buisson in Brazil; Sandra Díaz, Lucas Enrico, and Pedro Jaureguiberry in Argentina; John Silander and Joel Ratsirarson in Madagascar; Yadvinder Malhi and Anabelle Cardoso in Gabon; the Tropical Biology Association, Duncan Kimuyu, and Colin Beale for East Africa; Norman Owen-Smith, Izak Smit, Winston Trollope, Harry Biggs, and Rina Grant for South Africa; David Beerling, Andrew Illius, and Colin Osborne in the UK, along with the remarkable group of fire palaeoecologists including Andrew Scott, doyen and entertaining guide to the deep past and Claire Belcher, brilliant experimentalist on ancient ecosystems. In Europe, many thanks to Joris Cromsigt, Mariska te Beest, Steve Higgins, Herbert Prins, Simon Chamaille, and Herve Fritz. The MEDECOS group has been a training ground and inspiration for many of my generation: thanks to Jon Keeley, Juli Pausas, Phil Rundel, and to Byron Lamont and Tianhua He for their surge of phylogenetic research on the ancient history of fire. In recent times in the USA thanks for great field trips to Joe Craine, and Quinton and Elizabeth Martins, and to Lars Hedin and Simon Levin for a friendly introduction to east coast academe. In India, we had exciting times, and great food, with Jay Ratnam and Mahesh Sankaran, and in China, also with great food, with Tristan, Kyle Tomlinson, and our hosts at Xishuangbanna Tropical Botanical Garden.

Finally, I thank Elizabeth, Helen, and Christopher for their interest, encouragement, and support over the years. Winifred has been the rock on which all was built.

Contents

CHAPTER 1

Introduction to open ecosystems: a global anomaly and a local example

'The general, and almost entire, absence of trees in Banda Oriental is remarkable...It has been inferred with much probability, that the presence of woodlands is generally determined by the annual amount of moisture; yet in this province abundant and heavy rain falls during the winter....We see nearly the whole of Australia covered by lofty trees, yet that country possesses a far more arid climate. Hence we must look to some other and unknown cause'. (Charles Darwin, 1839, Voyage of the Beagle, Chapter 3).

1.1 Introduction

This book is about the non-forested open ecosystems of the world. In a recent analysis for conservation, Dinerstein et al. (2017) estimated that 41.5 per cent of the world's land area was covered by forested ecoregions, with the remaining 58.5 per cent being non-forested ecosystems (Table 1.1). The non-forested ecoregions include deserts too dry for forests, tundra too cold for forests, and soils that are flooded or saline and hostile to tree growth. What fascinates me, however, are the vast areas where climates are warm enough and wet enough for forests but are dominated, instead, by grasslands, shrublands, savannas, and open woodlands. These open ecosystems, covering more than a quarter of the world's land area, are the subject matter of this book. Ever since Humboldt's voyages to South America, biogeographers have puzzled over the origins of these open ecosystems (Wulf 2015). Their existence challenges the view that climate determines the major vegetation formations of the world—they should be forests. Tropical savannas are a particularly glaring anomaly, with vast areas occurring in climates that also support closed forests. But striking examples also occur in cooler regions at higher latitudes and elevations on all vegetated continents.

Open ecosystems (abbreviated as OEs in this book) are largely ignored in ecological textbooks. Forests represent the less disturbed ecosystems and have received far more attention in developing ecological concepts. In a survey of papers on tropical biodiversity published from 2004 to 2009, Bond and Parr (2010) noted a ratio of >8:1 papers on tropical forests versus the non-forested tropical biomes (1343:164). Since then, this bias has started to change and there has been a remarkable growth of research on open ecosystems and the processes that maintain them, especially in the tropics. This book is intended as a guide to the lighter side, the open, non-forested ecosystems of the world. Open ecosystems pose profound challenges to long-held concepts in ecology, biogeography, and paleoecology. You may also discover the peculiar cultural biases in many parts of the world that underlie ideas on what is 'natural' or 'healthy' in contrasting ecosystems. These cultural blind spots mean that there are still many fundamental discoveries to be made in the non-forested ecosystems of the world. The study of ecology and evolution on the light side is an exciting endeavour. Furthermore, there are profound implications for understanding global change impacts, Earth–atmosphere feedbacks, and

Open Ecosystems: ecology and evolution beyond the forest edge. William J. Bond, Oxford University Press (2019). © William J. Bond 2019.
DOI: 10.1093/oso/9780198812456.001.0001

Table 1.1 An estimate of the extent of forested and non-forested landcover from Dinerstein et al. (2017). Estimates of the extent of open ecosystems in climates suitable or unsuitable for forests are also shown based on interpretation of ecoregions listed in Dinerstein et al. Areas do not take into account transformed vegetation.

Biome types	10⁶ km²	%
Forested biomes	61.1	41.5
Non-forested biomes	86.1	58.5
Total vegetated	147.2	100
Non-forested biomes		
Warm enough and wet enough for forests		
Tropical & Subtropical Grasslands, Savannas	31.5	21.4
Temperate Grasslands, Savannas, & Shrublands	10.2	6.9
Mediterranean Forests, Woodlands, & Scrub	3.3	2.2
Montane Grasslands & Shrublands	4.9	3.3
Total		33.9
Too dry for forests		
Deserts	26.3	17.8
Too cold for forests		
Tundra	8.8	6
Soils unsuitable for forests		
Flooded Grasslands	1.2	0.8
Total non-forested	86.1	58.5

for developing policies on how to adapt to, and mitigate against, the effects of global change.

1.2 Fynbos shrublands: an example

In the natural areas around Cape Town, where I live, shade is a rare commodity. Natural forests do exist if you look hard enough but they are tiny. Generally you will have an uninterrupted view of the Cape mountains over the colourful cape fynbos shrublands (Figure 1.1).

1.2.1 Climate does not explain the absence of forests

The climate of the Cape, the south-west corner of Africa, is Mediterranean with the unusual combination of a cold, wet winter and a hot, dry summer. In conformity with the climate, the vegetation of areas with substantial winter rainfall differs strikingly from the rest of Africa. Instead of

forests, savannas, or grasslands, these regions are dominated by shrublands. Similar shrublands occur in the other winter rainfall regions of the world; chaparral in California, matorral in Chile and the Mediterranean Basin (including its north African fringes), and the sclerophyll shrublands of western and southern Australia (Keeley et al. 2012). Since the 1970s, there have been intensive studies on these Mediterranean-type shrublands exploring the hypothesis of convergent evolution in response to the distinctive climate (Cody & Mooney 1978). At the drier end of the rainfall gradient, the Cape region is dominated by arid shrublands—the succulent Karoo—succulent because of the large numbers of stem and leaf succulent shrubs. With increasing rainfall, the succulents drop out and small-leaved shrubs dominate ('renosterveld'). As rainfall increases, or on sandy soils, 'fynbos' shrublands occur with a mix of small-leaved evergreen shrubs, an overstorey of taller broadleaved shrubs (mostly Proteaceae), and reed-like graminoids, the leafless 'restios' (primarily Restionaceae) (Kruger & Taylor 1980; Cowling 1992). Fynbos is the dominant vegetation in the famous 'Cape Floristic Region', the world's richest temperate flora (Goldblatt & Manning 2002). It occurs in a temperate climate varying from arid, 250 mm p.a., to humid with >3000 mm p.a. Despite this enormous rainfall range, fynbos structure remains strikingly similar throughout (Figure 1.2). The overwhelming impression to a visitor is of a heathy shrubland whether you are hiking in the mountains, stumbling over limestone hills, or on the sandflats skirting the coastline.

Why is this vegetation shrubby? Is the lack of trees a consequence of a Mediterranean-type climate? Foresters have inadvertently run the experiments by planting out pine trees from the northern hemisphere, and eucalypts and acacia species from Australia, since the early twentieth century (Bennet & Kruger 2016). Seedlings died and plantations failed at the arid end of the gradient (in the succulent shrublands). Plantations were successful but often of poor quality in renosterveld, but the trees thrived in large areas of the fynbos. The plantations were established to augment timber from indigenous forests. These native forests of the Cape were too small to meet the needs of the growing human population. Forests generally occur in patches too small

Figure 1.1 Fynbos, an open shrubland ecosystem, dominates the landscapes of the south-western Cape of Africa. However closed forest patches do occur as shown in the foreground. Soils are derived from quartzites and sandstones and are very nutrient-poor.

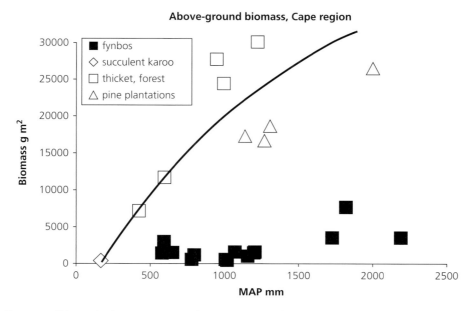

Figure 1.2 Above-ground biomass in relation to mean annual precipitation (MAP) for open shrublands versus closed forest and thicket. Biomass for pine plantations is also shown. Filled symbols are all from flammable vegetation that burns on decadal cycles. Open symbols are vegetation where fire is excluded. The line gives an indication of the climate-limited 'carrying capacity' for trees in the region. From Keeley et al. (2012).

to map at the national scale in South Africa (<0.5 per cent of the land area). Bowman (2000) described similar 'rainforest' patches in Australia as an 'archipelago' of island-like patches in a sea of sclerophyll woodlands, shrublands, and savannas. The Cape forests occur in a mosaic of forest patches in a matrix of fynbos. Forest boundaries are usually abrupt—just a few metres wide. Their existence, together with the plantations, indicates that the climate potential for growing trees is far greater than implied by the dominance of the low fynbos shrublands.

Figure 1.2 contrasts above-ground biomass of fynbos with that of indigenous forests and plantation forestry across a precipitation gradient. While the biomass of closed forests and thickets declines with increasing aridity, fynbos biomass remains remarkably constant in comparison. The huge gap between actual (fynbos) and potential (forest) biomass in higher rainfall regions demands explanation. Why is forest not more widespread in the region? Why does fynbos dominate? How can you tell when an open ecosystem is open because it is too dry or too cold for trees to grow, or is 'anomalously' open because the climate could support a closed forest? These are some of the central questions for studies on open ecosystems. The answers have been surprising, unsettling, and often controversial.

1.2.2 The dominance of fynbos is not a consequence of deforestation

By far the most popular explanation for the existence of non-forested, OEs is that they *were* forests but have been felled, burnt, browsed, and cleared for crops in a human wave of deforestation (e.g. Williams 2003). There is abundant historical support for the deforestation hypothesis of the origin of open habitat in Europe and eastern North America (though not without dissenters; Vera 2000). Unsurprisingly, the hypothesis of origins through deforestation has been widely invoked elsewhere where OEs are common, particularly in the tropics. However, it has seldom been invoked for Cape fynbos. Fynbos shrublands are a major component of the Cape Floristic Region, the richest temperate flora on Earth (Cowling et al. 1996). There are six endemic families, 160 endemic genera, and ~9000 species, nearly 70 per cent of which are endemic to the region (Goldblatt & Manning 2002). A feature of the flora is the presence of very speciose genera such as *Erica* (656 species), *Aspalathus* (Fabaceae, 272 species), *Pelargonium* (Geraniaceae, 148), *Agathosma* (Rutaceae, 143), *Phylica* (Rhamnaceae, 133), *Muraltia* (Polygalaceae, 106) and a host of geophytes including *Oxalis* (Oxalidaceae, 118), *Moraea* and *Gladiolus* (Iridaceae, 115 and 105), and *Disa* (Orchidaceae, 92). Many of the species have tiny distribution ranges such as *Mimetes stokoei*, a rare beautiful shrub in the Proteaceae spread over an area about the size of a tennis court. Fynbos species are notoriously intolerant of shade. Gardeners who plant popular Proteaceae species too close to a wall will never see them flower. Pine trees spreading from plantations cast sufficient shade to prevent flowering and ultimately kill the fynbos understorey (Richardson & van Wilgen 1986). This extreme intolerance of shade is characteristic of many plants growing in open ecosystems in many parts of the world.

The combination of extraordinarily high diversity of an extremely shade-intolerant flora indicates that fynbos is not cobbled together from remnant understorey plants of a mythical primeval forest felled by people. Instead there is evidence that fynbos lineages have evolved over millions of years throughout the Cenozoic (Linder 2005) with some common elements dating from the mid Cretaceous (He et al. 2016). How can such an ancient system have persisted for so long in a climate where forests can grow and forest trees can shade fynbos elements to extinction within a generation?

1.2.3 Forests are not excluded from fynbos by low nutrient soils

Soils hostile to tree growth are a favoured explanation for the absence of forests where the climate is otherwise suitable. In the 1950s, Beadle in Australia documented high soil phosphorus under closed forest and low soil P in sclerophyll woodland (Beadle 1954). The sclerophyll woodlands ('forests') have a fynbos-like understorey of shade-intolerant evergreen shrubs but with that uniquely Australian addition of tall eucalypt trees. The idea that low soil P accounts for open sclerophyll shrublands has since become part of the folklore of southern hemisphere ecology (e.g. Hopper 2009). Fynbos occurs on some of the most P-poor soils in the world (Cramer et al. 2014). The soils are derived from extremely infertile quartzites and sandstones in the mountains and dune sands in the lowlands. Fynbos also occurs on pockets of richer soils, derived from granites, shale, and limestone. All these geological substrates also support patches of closed forest or thicket. Forest soils are richer in nutrients than matched fynbos sites (Table 1.2), a very general result in comparisons of open versus closed ecosystems worldwide.

The key question here is whether the forests occur on nutrient rich soils because of high nutrient requirements, or whether the soils are richer in nutrients because of enrichment by the forest. Vegetation alters soils, most profoundly in the surface layers where ecologists usually sample them, a feedback of vegetation to soil chemistry that is too easily overlooked (Wigley et al. 2013). Sophisticated theory and experimental studies have explored factors controlling extraction of soil resources and their importance for plant competition (e.g. Tilman 1982; Craine 2009). However, the feedbacks of plants to those soil resources have been somewhat neglected in the conceptual framework. Yet there are many examples of how plants modify soils (see e.g. Jobbágy & Jackson 2001, 2004). In the case of fynbos, invasive trees have indicated the extent to, and the rate at, which these plants can alter very nutrient-poor sandy soils. Yelenik et al. (2004) studied soil

Table 1.2 Comparison of soil characteristics in adjacent fynbos and forest. Variables are ranked from greatest to least difference using Hedge's *g*, a measure of 'effect size'. Soils are derived from the same sandstone substrate. Data from Cramer (2010) for samples from Swartboskloof, Jonkershoek Valley.

Measure	Unit	Forest	Fynbos	*g*
Ca	cmol kg^{-1}	9.8±2.2	0.6±0.1	7.4
pH	in 1 M KCl	5.2±0.2	3.7±0.1	5.0
T-value	cmol kg^{-1}	15±2	5±1	3.4
Mg	cmol kg^{-1}	3.3±0.3	0.5±0.2	3.3
Bray II P	mg kg^{-1}	22.5±8.6	4.8±0.9	2.7
Total N	mg kg^{-1}	3.9±0.8	1.3±0.6	2.0
K	mg kg^{-1}	188±41	79±30	1.6
Na	cmol kg^{-1}	0.12±0.02	0.07±0.02	1.4
C	mg kg^{-1}	39±6	22±1	1.3
H$^+$	cmol kg^{-1}	1.4±0.4	3.2±0.9	−1.8

impacts of *Acacia longifolia*, a small leguminous tree from Australia, colonizing sandplain fynbos. The soils were among the most nutrient-poor in the fynbos region. Nitrogen return to the soil increased from 20 mg N/m^2/yr under fynbos to 170 under the acacia. In two or three decades, the acacia had more than doubled available inorganic N from 750 to >2000 mg N/m^2/yr and increased total N in the soil nearly tenfold from 0.1 to 0.9 mg N/g soil (Yelenik et al. 2004). The increased N stimulated a five-fold increase in biomass of a grass that invades cleared stands of the acacia. Clearly the nutrient status of this soil is far from fixed and can be radically altered by the kinds of plants that grow on it. For native vegetation, the nutrient differences between forests and fynbos, despite sharing soil derived from the same geological substrates, is just as likely to be a consequence as a cause of forest distribution (Coetsee et al. 2015; Cramer et al. 2019).

Plant impacts on soil properties have long been known in temperate and boreal regions where podsolization can profoundly alter a soil (Ovington 1953; Nihlgård 1971). The removal of oaks for shipbuilding was thought to have caused development of podsols by altering litter quality when conifers replaced the oaks. The potential for feedbacks of vegetation to soil properties is one of the great unknowns in explaining vegetation mosaics of closed and open ecosystems. In the case of fynbos, we must ask whether the many distinctive nutritional adaptations, sclerophylly, cluster roots, parasitism, and carnivory (Cramer et al. 2014, Lambers et al. 2008) are a response to low nutrient soils—or a cause. Are these nutritional strategies an example of niche construction? Are fynbos plants creating their own misery? Experimental studies suggest that the fynbos environment is hostile for forest tree seedlings and reduces the likelihood of forest invading fynbos (Manders 1990).

1.2.4 Fire and the origin and maintenance of open ecosystems

Like most open ecosystems in warmer, wetter climates, fynbos burns regularly. Fynbos typically burns from 10–30 years after the previous fire and stands are seldom older than 50 years. For most of the twentieth century, fires were attributed to irresponsible human agency and thought to be major causes of environmental degradation. The issue was considered so serious that the country's leading vegetation scientists met in the 1920s to discuss the national problem of 'veld burning'. The consensus then was that fires were destructive and that every effort should be made to reduce their frequency and, depending on resources, suppress them altogether. Compton, an eminent botanist, summed it up by stating that 'the burning of the scrub on the Cape mountains is, in the long run, entirely destructive and absolutely indefensible' (Compton 1926).

Botanists in the Cape were convinced that fires were a major threat to extinction of unique Cape plant species. Among these were *Orothamnus zeyheri*, the marsh rose, a beautiful member of the Proteaceae, and *Serruria florida*, the 'blushing bride', much coveted for bridal flower posies. Populations of both species were declining and major efforts were made to protect the plants and suppress fires. Disaster struck when a fire swept through the remaining *Orothamnus* populations in 1969. Based on best scientific knowledge of the time, the fire should have delivered the *coup de grace*. But, to the astonishment of the conservators, thousands of seedlings emerged from the scorched land producing vigorous healthy plants and renewing the marsh rose (Boucher 1975).

We now know that fire is essential in the life history of thousands of fynbos plant species. Fire stimulates flowering of geophytes, seed release of serotinous shrubs, or seed germination cued by heat or smoke (Bond & van Wilgen 1996). Plants are adapted to cope with fire by resprouting, producing thick bark that insulates stems, or evolving nonsprouting behaviour. The vegetative form itself may be subject to selection for increased flammability, for example through retention of dead leaves and branches (Bond & Midgley 1995; Schwilk & Ackerly 2001).

In contrast to fynbos, forest plants do not readily burn (Figure 1.3). Fires spreading from fynbos seldom penetrate more than a few metres into forest. The contrast in flammability is not due to differences in biomass, but to the arrangement of biomass and its effects on flammability (van Wilgen et al. 1990). Fynbos shrubs have finer branches and retain dead branches more often than forest species. Forest trees

and shrubs differ strongly from fynbos shrubs in having larger leaves, thicker stems, and little or no dead leaf retention. Forest trees also have very different pollination and dispersal modes from open ecosystem species and have no dependence on fire for recruitment. Thus both reproductive and vegetative traits are completely different in the open and closed ecosystems, a difference repeated in similar open versus closed comparisons around the world.

There has been an explosion of research on fire and vegetation in the last two decades. Studies have explored fire as a selective force shaping plant traits, as an agent (consumer) structuring vegetation, and as an evolutionary force shaping vegetation structure over the long history of plant life on land. These studies have swept away (scientific) prejudices against vegetation fires that marked the science of much of the twentieth century. The power of cultural bias is demonstrated by the fynbos example. How could highly trained botanists possibly fail to notice the abundant evidence for fire-adaptive traits in fynbos? How could they not explore fire as a key ecological process maintaining the fynbos system? But, before we get too critical, one wonders who will judge us for missing the obvious because of our cultural blinkers and what is it that we are missing?

Fire and large vertebrate herbivores are the two major classes of consumers of woody biomass that influence vegetation structure. Both have the potential to carve open forests, creating and maintaining light-saturated ecosystems instead. Fynbos forage quality is too poor to support significant herbivory so fire is the major consumer in the region. We will see that large vertebrates were (and sometimes still are) influential in opening up forests elsewhere in the world.

1.3 Open ecosystems in wider context

Cape fynbos, in a microcosm, raises some of the key questions explored in this book. What is the ecology, biogeography, and evolutionary history of open vegetation? Vegetation structure has received far less attention than productivity in global studies of ecosystems. But, arguably, structure is at least as, if not more, important than productivity for the habitats of plants and animals, their functional traits, and for ecosystem goods and services. Ecosystem

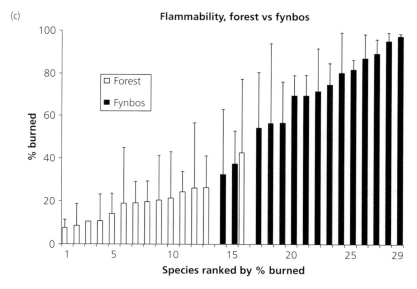

Figure 1.3 Pyrophilic fynbos and pyrophobic forest. A: A sharp boundary between burnt fynbos on the right and forest on the left a few hours after a fire had burnt the fynbos. B: Forest/fynbos mosaic after an intense fire. The fire did not penetrate beyond the forest margin. C: Contrasting flammability of forest (open) versus fynbos (closed bars). Flammability of whole shoots was measured as % sample branch burned in a standardized method. Plants that contribute to 'fuel load' in fynbos mostly have fine branches and retain highly flammable dead leaves and stems (ex. Burger & Bond 2015).

structure is also important for earth science. Grasslands differ from forests in albedo, the rate of rock weathering, and evaporative effects on climate (Beerling & Osborne 2006; Beerling et al. 2012). Conifer forests have a net warming effect in boreal regions because they shed snow, presenting a dark canopy with low albedo, unlike boreal shrublands where deep snow reflects sunlight back into space (Betts 2000).

In the Cape region, foresters began establishing plantations of conifers in the 1930s. Persistent droughts in the 1920s led to a controversy over whether more plantations should be planted to create a wetter climate or whether plantations were drying up the rivers (Bennett & Kruger 2016). Some of the earliest hydrological catchment experiments in the world were set up to answer these questions. By the 1970s the answers were clear—forests dried up the rivers (Figure 1.4). Legislation was passed that restricts new afforestation depending on downstream effects and competing demands for water. In the 1990s, a massive employment scheme was created to remove invasive trees from mountain catchments and ensure the streams and rivers

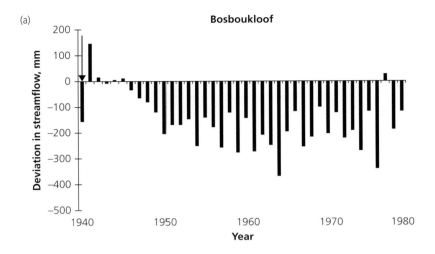

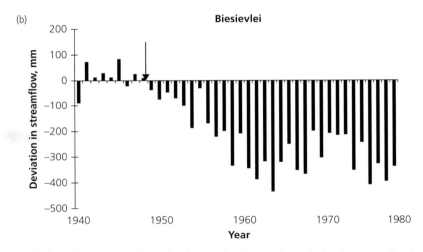

Figure 1.4 Conversion of fynbos to forest causes major declines in streamflow. Data are from paired catchment experiments where, after a calibration period, one catchment was planted up to conifers and the other left as fynbos. The year of afforestation is indicated by a vertical arrow. Results from two experimental catchments are shown, A (57% of the area afforested) and B (98% afforested). Data from van Wyk (1987).

continued to flow from the mountain catchments. Fynbos, it turns out, is an excellent vegetative cover transmitting 65 per cent of the rain into streamflow, in marked contrast to plantation forests that evapo-transpire rainfall before it enters the rivers. The results of these hydrological experiments contradicted long-held beliefs that planting trees ameliorated climates dating back, at least, to Humboldt's speculations in South America. Data such as these motivated a major programme to clear invasive trees, 'deforesting' the fynbos-vegetated catchment areas. The clearing programme yielded

additional benefits by creating jobs and protecting the rich shade-intolerant fynbos flora from shading. The hydrological trends have been repeated in many subsequent studies around the world with reduction of streamflow following afforestation (Bosch & Hewlett 1982; Farley et al. 2005). Yet public perceptions that forests ameliorate climate are still widely held and sometimes promoted by foresters.

Contrasting with the awareness of the negative effects of afforestation on water supply is the worldwide movement to plant trees to sequester carbon and stave off global warming. Among the greatest threats to the world's remaining natural forests is their conversion to open habitat, grasslands, and croplands, in the face of a growing tide of humanity. The threat to forests is well understood with considerable resources devoted to documenting forest losses and a parallel development of international policy to reduce deforestation and to support tree planting in deforested areas. Ecosystems on the light side are far less resourced or understood. Legislation and international initiatives are often antagonistic to maintenance of open ecosystems. Fires are essential for maintaining most open ecosystems but legislation for fire suppression to protect forests is very widespread. The consequences of loss of open ecosystems for the enormous diversity of plants and animals that dwell in the sunlight is little understood or appreciated. The cultural context, the idea that forests are nature's supreme expression, is common in Europe and parts of North America. But this notion is rare, or absent altogether, in those parts of the world where flammable OEs are the norm.

My hope is that this book might spread at least greater scientific awareness of the fascination and distinctive features of open ecosystems and their ecology and evolution.

1.4 Geographic scope and terminological difficulties.

This book is not an exhaustive review. Each of the chapters could be extended into a book on its own. Instead the book develops a point of view, an argument, with examples from different systems and from different parts of the world. Many of the examples come from places I have visited and experienced first-hand. The terminology for open ecosystems varies greatly from place to place and it is often extraordinarily difficult to understand whether it is a closed or open system from the literature. The word 'forest' is a classic example of something which can invoke entirely different images depending on the reader and the systems with which they are familiar. For example, I was astonished to discover that teak 'forests' in India are savannas with a

Figure 1.5 Giant *Eucalyptus regnans* in Tasmania. These are among the tallest trees in the world yet they cast very little shade. Despite high biomass, eucalypt 'forests' are open ecosystems where they support a shade-intolerant understorey. In this example, the eucalypts recruit after very infrequent fires and the understorey is made up of closed 'rainforest' species (see Tng et al. 2012).

C_4 grassy understorey when not invaded by *Lantana* or *Chromolaena*. Longleaf pine 'forests' in the southern USA are also savannas and have many attributes like the savannas of Africa, Australia, and South America. To help clarify the terminology, Ratnam et al. (2011) gave a functional definition of a 'forest' and how it differs from a 'savanna'. This functional definition is what is used in this book to distinguish 'open' from 'closed' ecosystems. A key field criterion is whether the understorey (ground layer of plants) is shade-tolerant or shade-intolerant. If the former, it is a 'closed' forest. If the understorey is shade-intolerant it is called an 'open' forest. In many parts of the world, open ecosystems are also low biomass systems because of the scarcity of trees. However, open ecosystems include some very high biomass ecosystems. For example, it is difficult to avoid the term 'forest' for the giant eucalypts of Australia, which include the tallest angiosperm trees in the world. Yet the understorey below these giant trees is often of shade-intolerant woody shrubs and graminoids. They are forests but 'open forests' by my usage of the term (Figure 1.5).

1.5 History of research

The problem of open vegetation in climates that could support forests has been recognized for more than a century in various parts of the world (Darwin 1840; Grandidier 1898). Especially in the tropics and subtropics, scientists from Europe tried to make sense of the new worlds they were observing, inevitably using the concepts from their own temperate world. The key idea of climate determining the distribution of the major vegetation formations first emerged from Humboldt's explorations of montane layering in the tropics (Wulf 2015). Since Humboldt, forests had been elevated to 'climax' conditions, the supreme expression of nature, wherever the climate is warm enough and wet enough. Departures from the forested condition were, and still are, widely seen as due to human activities, especially deforestation using fire or directly by felling and cultivation of crops. This hypothesis has been assumed true for two centuries and is still very widely believed by both scientists and the public, for reasons other than straight scientific understanding (Pausas & Bond, in press).

1.5.1 Consumer control: fire and open ecosystems

The power of fire to create and maintain open vegetation, regionally and globally, has been known since the first half of the twentieth century (e.g. Phillips 1936; Sauer 1950). Sauer wrote a coherent account of global fire regimes and their importance in shaping vegetation in the 1950s. Experimental studies manipulating fire regimes were initiated by French and British ecologists from the first half of the twentieth century to test whether regular fires could, indeed, account for the absence of forests in the tropics. Some of these experimental studies are continuing today. Impressive reviews on the ecology of fire appeared more than 50 years ago, and the study of fire ecology was greatly enhanced by Ed Komarek's initiation of the Tall Timbers Fire Ecology conferences in Florida in the USA.

1.5.2 Consumer control: herbivores and open ecosystems

Unlike fire, herbivory has long been studied and long been part of mainstream ecology, as reflected in chapters in general ecology textbooks. In this book, I only discuss vertebrate herbivory as regulator of tree populations, potentially creating and maintaining open ecosystems. Insect herbivory is often specialized on a particular tree species whereas vertebrate herbivory is not. The role of vertebrates in shaping ecosystem structure has been frequently studied, but there seem to be no comparable studies on the role of insect herbivory in preventing forests from forming.

Herbivory by the mammalian megafauna has been much studied in the context of rangeland and wildlife ecology. Megafauna, in this book, refers to the collective set of all larger mammal species in a fauna (usually >30 kg). The term 'megaherbivore' refers to those very few species with body mass greater than 1000 kg. These giant beasts have been identified as qualitatively distinct in a number of their characteristics (Owen-Smith 1988). They are generally too large for predators to take sufficient numbers of prey to regulate megaherbivore populations. Consequently their populations are likely to grow until limited by food or other critical resource, making them particularly influential in shaping the ecosystem. Their large size also means they have large effects through consuming large quantities of food, and through direct physical impacts on vegetation. Megaherbivores suffered differentially high extinction rates in the Pleistocene across the world, except Africa. The ecological effects of these extinctions have become a topic of great interest over the last few years (Johnson 2006; Svenning et al. 2016). African ecosystems are the last to give us glimpses of the complex ecology of ecosystems with the full range of herbivores of different sizes, with multiple predator species, long human occupancy, and regular fire.

The influence of mammal herbivory on OEs is not confined to megaherbivores. In many open ecosystems, trees are excluded by browsing herds of deer, antelope, caprids (goats, sheep), equids, bovids, and other large herbivores. Deer 'overabundance' for example, is changing the structure of temperate and boreal forests, converting them to open, savanna-like ecosystems (Côté et al. 2004). Both herbivory and fire can be viewed as 'consumers' (Bond 2005). Both remove plant biomass and convert it to complex organic substances. Both are sufficiently regular and predictable to select for plant traits adapted to the pattern of consumption. These plant traits, in turn, can feedback to the consumer, altering fire properties, or the number and kinds of herbivores. In these feedbacks, they differ from purely physical disturbance such as landslides, earthquakes, cyclones, and the like. Such disturbances may select for plant traits but these do not feedback to influence the frequency or magnitude of the physical disturbance. Consumer control (by fire or large vertebrate herbivory) is a major theme of this book.

1.5.3 New developments

Given that a lot has been known about open vegetation for many years, what has changed? The scale of analysis is the most obvious. Debates over the origins and maintenance of open and closed ecosystems were typically regional in nature. Each country has its own literature trying to explain the anomaly of open vegetation. Earth observation data has allowed the scale to change from local to regional, continental, and now global. And with that change in scale came the recognition that the

problem of open ecosystems is a general one, with answers sought using models, experiments, and observations from studies from long-neglected ecosystems from around the world. Along with the change in spatial scale, there have been advances in the exploration of the past, especially the deep past. Fossils and phylogenies have revealed the great antiquity of fire (400+ Ma) as a major process structuring vegetation. Fires have been burning for most of the history of plant life on land. This was a radical advance given that most biologists in the twentieth century assumed fires were a human artefact degrading environments. The constructive force of fire had now to be recognized as operating from the deep past to the present and into our uncertain future. It is interesting to note, for example, that fire was not even mentioned in the influential hypothesis of Hairston, Smith, and Slobodkin (1960) on how trophic interactions can mould ecosystems. Even in the twenty-first century, White (2005), an Australian from that most flammable of continents, could conclude that nitrogen controlled world vegetation by limiting animal herbivory, without considering fire as a major agent of change. But it has now clearly emerged that fire is at least as influential as vertebrate herbivores in altering vegetation structure, composition, and functional traits both in modern ecosystems and in those of the deep past. So, the recognition of the antiquity of open vegetation is radically changing perceptions of these systems.

The tools used for recognizing their antiquity are also new so that technological advances have contributed greatly to our changing understanding of the forces controlling nature. These tools include stable isotopes, charcoal and other proxies for fire, and molecular phylogenies which supplement fossil studies of the past. Along with the markers of open vegetation and the processes maintaining them is the development of a conceptual framework that helps explain open ecosystems while at the same time displacing the successional framework developed early in the twentieth century. Thus, where grasslands were seen as early successional stages on a successional trajectory to the climax state of forest, we may now view forests and grasslands as alternative stable states. We need no longer blush when talking about 'early successional' grasslands that have remained 'early successional' for thousands, or sometimes hundreds of thousands of years. We can replace the inappropriate vocabulary of succession with the jargon of alternative states maintained by positive feedbacks in each state that prevent transitions to the alternative state. This new conceptual framework has the liberating result of forcing us to recognize the profound role that organisms play in creating their own environmental conditions. Far from being passive responders to, say, climate or soils, plants can alter their environment by altering the distribution of nutrients in the soil or the light received by seedlings. Of course, we have known about most of these feedbacks for many years. Podsols, for example, were known to develop from brown earths when oaks were replaced by conifers in northern ecosystems. What was missing was the scale of these changes, and their importance in determining world vegetation patterns.

1.6 Book structure

This book explores the profound changes in understanding of open vegetation and the processes which account for their openness. In doing so, we will discover new insights into fire as a global process (Chapter 7) which, along with large vertebrate herbivores (Chapter 8), has maintained open vegetation for millions of years (Chapter 5). We will explore the extraordinary diversity of open ecosystems (Chapter 4), the evolutionary history of this diversity, and its importance as evidence for the antiquity of non-forested ecosystems (Chapter 5). We will consider how soils can influence forest distribution through their physical and chemical effects on tree growth (Chapter 6). Intrinsically hostile soils for tree growth may have provided the cradle for the evolution of OE plants during times of high forest cover. Looking to the future, where vegetation is not at equilibrium with climate, clearly climate change will have more indirect effects on the ecosystem. We will see how changes in atmospheric composition, including increasing carbon dioxide, may profoundly alter open ecosystems and threaten their future (Chapter 9). Finally, we will consider how OEs are managed, plans to replace them with forests to 'restore' them to a supposed pristine condition, and their potential contribution to geoengineering a future cooler world (Chapter 9).

CHAPTER 2

The pattern of open ecosystems and the climates in which they occur

2.1 Introduction

Climate has long been considered the prime factor determining the distribution of major vegetation formations, implying that there is a single stable vegetation state for a given climate. Thus tropical forests are characteristic of climates that are warm and wet, deserts where climates are dry, boreal conifer forests where it is too cold for broadleaved deciduous trees. But the assumption of 'one climate = one vegetation' is not true for large parts of the world where strikingly different vegetation states, such as forests and grasslands, occur in the same landscapes sharing the same climate. For most of the twentieth century the non-forested ecosystems ('open' ecosystems) were thought to be secondary vegetation produced by deforestation and anthropogenic burning. While there is no question that deforestation has occurred, and is increasing at an alarming rate, there is also growing uncertainty as to the anthropogenic origin and extent of 'deforested' ecosystems.

Some of the world's richest biodiversity hotspots are open ecosystems, implying they have been stable and persistent long enough for substantial speciation to have occurred. Quaternary studies are showing that, in some regions, they were even more extensive before the rise of human populations. And, in deep time, fossil and phylogenetic evidence indicates that open ecosystems have ancient origins while the species they contain date back millions of years.

Recognition of the age and extent of open ecosystems challenges traditional ecological concepts. For example, succession, driven by competition for resources and especially the fight for light, should result in dominance of taller, shade-tolerant growth forms where resources permit. Yet in climates that support forests, open vegetation dominated by low-growing plants is also widespread and has apparently persisted for millions of years. In this chapter I explore the global extent of the mismatch between actual and potential vegetation and characterize the climate conditions in which these ecosystems most often occur.

2.2 Climate explanations for the distribution of biomes

The first attempt to classify and map world vegetation and to document links with climate began with Alexander von Humboldt's travels in the late eighteenth and early nineteenth century. Early attempts to classify and map vegetation attempted to include information on both taxonomic (species, family, genera) and structural/functional attributes (see Keddy 2007; Shugart & Woodward 2010 for accessible accounts). The first world-scale vegetation map was produced by Grisebach in 1872, in which he also tried to combine floristic and structural information. The first modern vegetation map of the world was produced by Schimper (1903). Schimper emphasized structural similarities in vegetation occurring at similar latitudes and sharing similar climates. Regardless of dissimilarities in floristic composition, vegetation converged to similar physiognomy where climates were similar. For example, regions with Mediterranean-type climates of winter wet, summer dry conditions were dominated by shrubs with evergreen leathery leaves even though the

Open Ecosystems: ecology and evolution beyond the forest edge. William J. Bond, Oxford University Press (2019). © William J. Bond 2019.
DOI: 10.1093/oso/9780198812456.001.0001

families and genera of the dominant shrubs are very different on different continents. The broad structural units identified by Schimper have persisted into modern usage as world biomes. They include Tropical Rainforest, Monsoon Forest, Summergreen Broadleaf Forest, Needleleaf Forest, Sclerophyll Woods, Savannas, Steppes, Heaths, Dry Deserts, and Tundra.

Climate–vegetation correlations began to emerge in the nineteenth century. Since climate at that time was poorly characterized, vegetation was widely used to identify major climate zones. The Koppen climate system was designed to align with ecological boundaries between major vegetation types and climates (Cramer & Leemans 1993; Shugart & Woodward 2010). The parallel development of vegetation classification and mapping and climatic classification and mapping led to persistent attempts to define biomes that were consistent with climate zones, and vice versa. It is, perhaps, this historical legacy that has led to the tautology where major vegetation types are assumed to be determined by climate and major climate zones are characterized by major vegetation types.

2.2.1 Examples of climate–vegetation correlations

Correlative climate–vegetation models were the first tools used for assessing how global change might impact on world vegetation. For example, Holdridge (1947) developed a triangular system where different vegetation units are predicted from three climate axes (although only two are needed) (Figure 2.1; Plate 1). Each vegetation unit is associated with its unique climate cell, its 'climate envelope'. As you might expect, the system works well in generating a potential world vegetation map consistent with mapped vegetation from twentieth century climate information (Cramer & Leemans 1993). To project future world vegetation, GCM projections of future climates have been used to predict the future distribution of vegetation types. Climate envelope models (= Species Distribution Models, SDMs) have since become very widely used to predict climate change impacts on species distribution (Midgley et al. 2002; Thuiller et al. 2005, 2008). They have also been used to assess biome responses to climate change using a wider array of climate variables fitted with

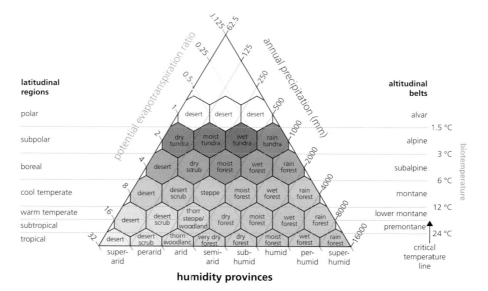

Figure 2.1 The Holdridge system. Each vegetation unit is uniquely associated with a climate envelope defined by two of biotemperature, precipitation, and potential evapotranspiration ratio. Biotemperature is a heat summation index calculated from average temperatures with the substitution of zero for all intervals where the temperature is below 0 °C or above 30 °C. Precipitation is the mean annual precipitation (mm). Potential evaporation is mm of water evaporated or transpired calculated as 58.93 times the biotemperature. The potential evapotranspiration ratio is the ratio of potential evaporation to annual precipitation. Values >1 imply humid climates while values <1 indicate more arid climates. (See Plate 1 for colour version.)

increasingly sophisticated statistical models (Heik-kinen et al. 2006; Elith & Leathwick 2009).

A related approach focusses on the major growth forms that constitute biomes, rather than the biomes themselves. Major biomes are characterized by a mix of characteristic plant types. Following the one-climate/one-vegetation type assumption, it is logical to expect that the distribution of plant types is also determined by climate. Eugene Box developed an intricate system where some 150 plant types were each associated with particular climatic conditions (Box 1981, 1996). Climate change should shuffle the plant types, depending on their climate preference, producing new vegetation patterns. This system has been applied by Bergengren et al. (2011) to assess how climate change will impact 'ecological integrity' (the distribution of major vegetation types) in the future. They identified regions most and least vulnerable to ecological change as a result of climate change. Boreal regions were the most vulnerable, according to their analysis, and the tropics, particularly tropical savannas, among the least vulnerable to climate change. Process-based models come to very different conclusions (see Chapter 9).

2.3 Climate zones where ecosystems are uncertain

The Holdridge, Box, and similar systems for predicting vegetation from climate all assume that, at equilibrium, one climate = one vegetation type. Whittaker (1975) also used climate correlations to locate major biomes on a temperature/precipitation plane (Figure 2.2). However, his system was unusual in that it recognized a climate envelope where the expected ecosystems were uncertain—that is where two or more stable vegetation types co-occurred. He described this climate envelope as including '… a wide range of environments in which either grassland, or one of the types dominated by woody plants, may form the prevailing vegetation in different areas' (Whittaker 1975, p.65). The major vegetation types in this envelope can have strikingly different physiognomies: tall closed forests, grasslands, shrublands, parklands with scattered trees, etc. are all found and often occur as mosaics in the same landscape. The climate zone where ecosystems are uncertain (EUCZ = Ecosystem Uncertain Climate Zone) is in the warmer areas of the world (MAT 10–30 °C typical

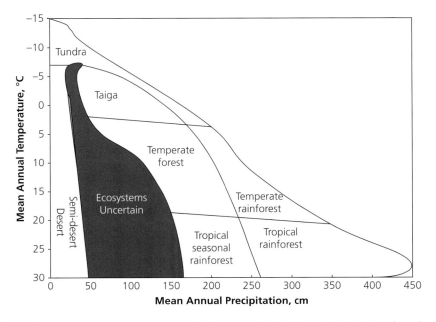

Figure 2.2 Whittaker's climate envelopes for major world vegetation formations. The shaded area is the climate envelope where ecosystems are uncertain and 'either grassland, or one of the types dominated by woody plants, may form the prevailing vegetation in different areas' (redrawn from Whittaker 1975, p. 65).

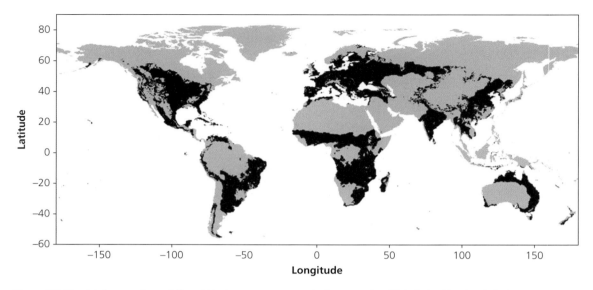

Figure 2.3 The global extent of areas falling within the ecosystems uncertain climate envelope (dark shading) in Whittaker's climate/vegetation system. From Bond (2005).

of low to mid-latitudes) and neither arid nor humid (~500–1600 mm per annum).

The global distribution of the EUCZ is shown in Figure 2.3 (Bond 2005). It is vast. Textbooks reproducing Whittaker's climate diagram for biomes sometimes label the EUCZ as 'grassland' or 'savanna'. And indeed, tropical and subtropical forests and savannas are major parts of this climate zone. However the EUCZ also includes regions with broadleaved temperate forests, some conifer forests, grasslands, shrublands (including Mediterranean type shrublands), the European steppes, moorlands, heathlands, and other diverse vegetation types. Very often strikingly different vegetation occurs as a mosaic in the same landscape in this climate zone (Figure 2.4; Plates 2 and 3).

Whittaker's (1975) climate envelope for uncertain ecosystems identifies which parts of the world can support very different vegetation within the same climate. It was the first attempt to identify climates in which open grasslands, shrublands, savannas, and woodlands co-occur with closed forests. Of course, the EUCZ does not tell us which of the possible vegetation types is most likely to occur in each region nor where alternative types form chequerboard mosaics. Rapid developments in remote sensing are allowing us to detect major vegetation types from space, including altered landscapes such as croplands

and human settlements. Continental, and indeed global scale, analyses of tree cover are now possible. What can these new tools tell us about the relative proportions of forest and open vegetation within the EUCZ and elsewhere in the world?

2.4 Tropical and subtropical regions

In the tropics, it seems reasonable to expect that tree biomass should vary linearly with increasing rainfall. Contrary to this expectation, tree cover does not gradually increase from dry to wet climates. Instead, tree cover is distributed tri-modally: sparse to absent in grasslands, present at intermediate levels in savannas, but dense in forests (Sankaran et al. 2005; Staver et al. 2011a; Hirota et al. 2011; Lehmann et al. 2011; Xu et al. 2016) (Figure 2.5). So this is not a continuum at all. Instead, tropical vegetation seems to form discrete communities of forest, savanna or grasslands (Dantas et al. 2016; Xu et al. 2016).

Although satellites allow us to see the planet as a whole, there are still difficulties in identifying closed-canopy forest from open-canopy vegetation from space. A closed forest casts too much shade to support the C_4 grasses characteristic of savannas. Using remote sensing to identify the presence of a grassy understorey is challenging: in savanna woodlands,

Figure 2.4 Examples of uncertain ecosystems. a) Mosaic of subtropical grassland, savanna, and forest in Gabon, Africa; b) Goias, Brazil; c) subtropical lowland forest/savanna mosaic, South Africa; d) mosaic of forest and savanna, Kakadu, Australia; e) montane savanna/forest mosaics, Drakensberg, South Africa; f) forest/ fynbos shrubland mosaic, Cape, South Africa; g) mosaic of oak parkland and mixed forest, Sonoma, California; h) Scottish moorlands with conifer plantation (Plate 2). (i) *Quercus robur* forest with *Festuca* grasslands, Carpathian Basin, Hungary (photo Á. Molnár). (j) *Betula pendula–Quercus robur* forest and *Stipa* steppe, Tula region, Russia (photo Yu. A. Semenishchenkov). (k) *Quercus pubescens* forest and *Stipa* steppe, Crimean Peninsula (photo Y. P. Didukh). (l) *Betula pendula* and *Festuca–Stipa* grasslands, Kostanay Region, Kazakhstan (photo Z. Bátori). (m) *Betula platyphylla* trees in *Leymus–Filifolium* grassland, Ulan Buton, Inner Mongolia, China (photo H. Liu). (n) Forest/steppe landscape with *Betula platyphylla* and *Stipa* grassland, Greater Khingan Range, China (photo H. Liu). (o) *Quercus brantii* woodland with *Bromus* grassland, Zagros Mts, Iran (photo A. Daneshi). (p) Mosaic of *Picea schrenkiana* forests and *Stipa* steppes, Xinjiang Uygur Region, China (photo H. Liu) (Plate 3). From Erdős et al. (2018), *Applied Vegetation Science*, 21(3), 345–62. (CC by 4.0).

Figure 2.4 Continued

it is difficult to detect grasses separately from the tree canopy layer. Identification of vegetation type is also scale dependent. For example, the MODIS tree cover product has a minimum resolution of 300 × 300 m, which is too large to detect the small forest patches which often occur in mesic savannas. For these reasons, Lehmann et al. (2011) used vegetation maps based on ground surveys to identify savanna regions in their comparative study of the physical environment of this biome in Brazil, Africa, and Australia. They found that precipitation was the main climate variable influencing savanna distribution. However, across the entire rainfall gradient, savannas never occupied the whole climate region (Figure 2.6). There

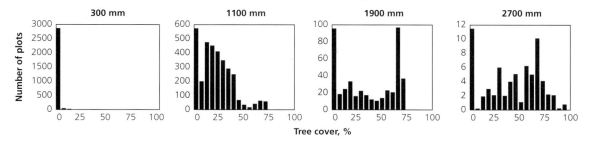

Figure 2.5 Histograms of % tree cover in sub-Saharan Africa categorized by rainfall at a 1 km scale. Each plot is of data from a 200 mm range of rainfall labelled in the mid-point of the range. Redrawn from Staver et al. (2011b). Percentage tree cover was derived from MODIS satellite reflectance data.

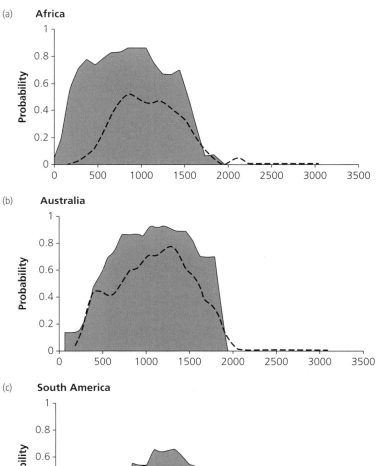

(a) **Africa**

(b) **Australia**

(c) **South America**

Figure 2.6 The probability of savanna occurrence derived from an analysis of vegetation maps across continental rainfall gradients of a) Africa, b) Australia, and c) South America. The dashed lines indicate the probability of fire. Note that on all three continents the probability of savanna occurrence is never 100% at any position along the rainfall gradient. Non-savannas lack grass in the understorey and, at higher rainfall, are mostly closed forests. Redrawn from Lehmann et al. (2011).

were always closed-canopy non-grassy vegetation types, shrublands at the arid end of the gradient but typically closed forests in humid regions (from as low as 650 mm MAP; Bowman 2000; Sankaran et al. 2005).

2.5 Temperate and boreal regions

2.5.1 Local studies

Anomalous occurrences of open vegetation in climates that can support forests are not restricted to tropical regions. Puzzling instances of mosaics of forest/grassland or forest/shrublands, not readily explained by human influence, are also common in temperate and boreal regions. Field studies have explored possible reasons for the contrasting vegetation types, concentrating especially on soil differences that might account for major vegetation structural changes. For example, Wells (1962) explored the complex pattern of grasslands, oak savannas, chaparral, and evergreen forests in central California. These strikingly different vegetation types form intricate mosaics in the hills and valleys of the region. Wells concluded that there was no strong correlation between geological substrate, soils, and the different physiognomic types. He noted that 'all physiognomic types sometimes occurred in close proximity on the same substratum and on the same slope' negating climate and geology as single factor explanations. He suggested, instead, that the elements of the mosaic were separated by different fire regimes, interacting with the substrate which influenced rates of recovery, and together selectively filtering compatible growth forms into the different vegetation types. Odion et al (2010) reached similar conclusions for mosaics of conifer forest and sclerophyll shrublands in the Klamath mountains of northwestern California. They argued that the shrublands are more pyrogenic than the forests, leading to development of fire-mediated alternative states. Jackson (1968), in Tasmania, explored the complex mosaic of sclerophyll shrublands ('button-grass'), eucalypt woodlands, and closed rainforest in the south-west of the island. Jackson argued that each vegetation type had a distinct fire regime which filtered the growth form mix while also feeding back to the fuel properties which contributed to the fire regime. Different physiognomic types also had add-

itional feedbacks to soil properties (Bowman & Wood 2009; Wood & Bowman 2012; Chapter 7).

Fire is not always the driver of open vegetation. Temperate grasslands, with little or no fire, also occur as anomalous patches in temperate forests, often on hilltops where they are called 'balds'. Weigl and Knowles (2014) reviewed balds in the mountain ecosystems of the Southern Appalachians in the USA, the East Carpathian poloninas in eastern Europe, and Oregon Coast Range grasslands of the western USA. The grasslands are apparently not the result of human forest clearance. Weigl and Knowles suggested they are legacies of the last glacial climate and megafauna, persisting today in cooler climates where tree growth is slow and where browsing by deer prevents forest recruitment (Weigl & Knowles 2014).

Vera (2000; 2002; Vera et al. 2006) controversially argued that western Europe had natural mosaics of open wood-pastures and closed broad-leaved forest. His idea challenged the widely held view that temperate Europe, in the absence of human influence, was covered throughout by a closed-canopy broad-leaved forest. Vera argued, instead, that heavy herbivory by herbivores that survived the end-Pleistocene extinctions had maintained an open pasture with scattered trees. Herbivores such as the extinct aurochs (ancestor of European cattle), tarpan (an extinct horse), bison, moose, and red and roe deer created and maintained open pastures. Trees such as shade-intolerant oak species were able to recruit into the wood-pastures under the protection of spiny shrubs in bush clumps (Bakker et al. 2004; Vera et al. 2006). This hypothesis of ancient open ecosystems in the European wildwood sparked intense interest and re-evaluation of paleo-ecological proxies for open and closed vegetation (e.g. Svenning 2002; Mitchell 2005). It also provided impetus to programmes of 'rewilding', the experimental re-introduction of large mammal herbivores into Atlantic ecosystems, and renewed interest in the conservation of 'semi-natural' grasslands in western Europe (e.g. Vera 2009).

2.5.2 Mapping the spatial extent of temperate and boreal open ecosystems

To help assess the wider distribution of open ecosystems in temperate regions, vegetation maps are a useful resource (cf. Lehmann et al. 2011). The classification and mapping of world vegetation as 'ecoregions' is based on such maps (Olson et al. 2001; Dinerstein et al. 2017). Table 2.1 lists the biomes and the geographical location ('realms') of temperate and boreal open-canopy vegetation derived from the ecoregions classification. Areas too cold for forests (tundra), or too dry (deserts) were excluded. The obvious candidates for uncertain ecosystems are ecoregions with mosaics of forest and grassland/shrubland. However, open vegetation living in climates with the potential for forming forests is also included. According to this analysis of the latest ecoregion data (from Dinerstein et al. 2017), temperate and montane open ecosystems account for around 15 per cent of the world's vegetated land surface. More than half of this estimated area is in the Palearctic (Eurasia). Forest/steppe mosaics, alone, are estimated to cover ~1.49 million km^2 (Table 2.1). A recent independent review of the vast Eurasian forest–steppe region estimated the area of this 'biome' as ~2.9 million km^2 or double that of the ecoregions estimate. It extends 9000 km east-west from 16°E (near Vienna, Austria) to 139°E in the Amur lowlands of Russia (Erdős et al. 2018). The next largest open ecosystem region is the montane grassland and shrubland biome also in the Palearctic. It includes the *Kobresia pygmaea* ecosystem of the Tibetan highlands, the largest montane grassland system in the world but growing in climates that also support forests (Miehe et al. 2019). Nearctic (North American) temperate open ecosystems are mostly grasslands, and include extensive C_4 dominated prairies, and oak and pine savannas which are similar to tropical and subtropical systems in their function.

Boreal regions also support extensive mosaics of forests and open shrublands and grasslands—the boreal forest/taiga biome in the ecoregions classification. The biome covers an estimated 15.4 million km^2, or 11.7 per cent of the world's vegetated land surface. About 10 million km^2 is in the Palearctic and 5.4 million km^2 in the Nearctic. For comparison, the entire land area of Australia is 7.7 million km^2.

Mediterranean-climate regions also have 'temperate' climates. Though they have been intensively studied, Mediterranean climate regions make up only a small fraction of world land area (Table 2.1). Shrublands often dominate these regions and the shrub growth form has long been viewed as an adaptation to the rigours of a winter-wet/summer-dry climate. However, closed forests also occur in Mediterranean-climate regions (e.g. fynbos and forest, Chapter 1) and here too, ecosystems are uncertain, with the shrublands often occurring in sites with the potential to support forests (Odion et al. 2010; Keeley et al 2012; Rundel et al. 2016).

2.5.3 Remote sensing of temperate and boreal trees

Availability of satellite imagery has opened up large-scale analyses of tropical forest and savannas (Bucini & Hanan 2007). Tree cover, biomass, and tree height

Table 2.1 The extent of open vegetation in temperate and montane regions based on estimates from ecoregions (ex Dinerstein et al. 2017). Values are area × 1000 km^2. Palaearctic = Eurasia+; Nearctic = North America; Afrotropic = Africa south of the Sahara; Neotropic = South America+. GL = Grasslands, SL = Shrublands.

| Realm | Biome | | | | |
	Forest/steppe mosaics	Temperate GL, savannas, SL	Montane GL, SL	Mediterranean ecosystems	Total
Palaearctic	1489	4775	3385	2029	11 678
Nearctic	136	3215		127	3478
Afrotropic			545	118	663
Australasia		583	68	844	1495
Neotropic		1622	871	148	2641
Total	1625	10 195	4869	3266	19 955

as measured from satellite imagery has indicated the vast extent of open vegetation in the tropics. It has also provided novel support for the existence of multiple attractors and multiple stable states of tropical tree communities (Staver et al. 2011a,b; Hirota et al. 2011). Scheffer et al. (2012b) were the first to apply similar analyses to northern vegetation. They analysed tree cover in boreal regions and found evidence for multi-modal tree frequencies suggesting multiple stable states not unlike the tropics. However, the different communities are arranged on a gradient of mean mid-summer temperature, not precipitation, consistent with the cold climate.

Subsequent studies have explored tree cover and tree height in north temperate and boreal regions. These have shown clear evidence for peaks and troughs in tree population attributes similar to those of the tropics (Xu et al. 2015, 2016; Abis & Brovkin 2017). With the availability of laser-based measures of tree height (lidar) large-scale analyses of tree height have been made (Xu et al. 2016). These have shown binary patterns of height, with tall trees in forests (high density tree cover) and smaller trees ('trubs') in more open habitat. If this trend is supported in subsequent studies, it would be consistent with tropical studies which show a similar pattern.

Satellite imagery in boreal, temperate, and tropical regions is revealing a tendency for trees to form closed forest, open savanna-like vegetation, or treeless grasslands/shrublands. The vast extent of open-canopy ecosystems in climates that also support closed forests is a glaring anomaly for the notion that climate determines the nature and distribution of major vegetation formations. Of course, the extent to which these analyses are confounded by human activities needs further analysis. However, the pattern of open and closed vegetation under the same climate also exists in sparsely populated regions such as the eastern forest-steppes of Eurasia (Erdős et al. 2018).

2.6 Beyond correlations

The title of Schimper's (1903) encyclopaedic book on world vegetation was 'Plant Geography upon a Physiological Basis'. His intention was to move beyond the well-established climate vegetation correlations and to describe the physiological properties of plants characteristic of particular climate regions. Walter (1973), in his book on the vegetation of the Earth, also emphasized physiological aspects of plants growing in different climatic regions. Subsequently, many studies have been conducted on the physiology of plants growing in different environments. These studies have explored how different plant forms and plant traits function in diverse environmental conditions. The motivation for many such studies is exploring the idea that the physiology of the organism is optimally fitted to the physical environment (e.g. Cody & Mooney 1977).

Many ecophysiological studies are primarily descriptive in nature. Woodward (1987) was among the first to suggest that physiology could be used in a predictive way to test whether climate determines major vegetation patterns. The apparent importance of climate in determining vegetation is a hypothesis and hypotheses need to be tested. This is no easy task. As Woodward (1987) noted, the problem in testing physiological hypotheses is that the object of study is 'vegetation, not…individual species, and vegetation is an unwieldy subject for experimentation'. The most tractable way forward was to build models simulating ecosystems using physiological principles. To do so would require detailed (daily or monthly) climate data to drive the physiological processes and soil physical properties (depth and texture) which modify plant available moisture. The properties of the simulated ecosystems could then be compared to those of actual vegetation. Large mismatches would imply inadequate physiological understanding or, perhaps, factors extraneous to physiology that are ecologically important.

2.6.1 Dynamic Global Vegetation Models (DGVMs and DVMs)

The first models capable of simulating global vegetation, including the major growth forms in different regions, began to emerge by the mid-1990s. They were called Dynamic Global Vegetation Models or DGVMs. Where the focus is not global, DVM is the acronym now used. DGVMs are complex models analogous to global climate models and initially developed to support them by providing better simulations of the changing vegetated land surface (Cramer et al. 2001; Shugart & Woodward 2010). The models simulate plant growth using daily or monthly input of climate variables. Unlike correlative models, they also simulate effects of changing atmospheric CO_2. DGVMs simulate vegetation in pixels using soil texture and depth from global soil maps. They have a modular design with, for example, a vegetation physiology module simulating photosynthesis and transpiration at minute to hour time steps. Carbon gained is allocated into different plant functional types (PFTs) which compete in a vegetation dynamics module at time steps of months to years. The assembly of dominant PFTs in a pixel effectively simulates 'biomes', allowing DGVMs to be used as biogeographic tools. The first DGVMs had a very limited set of PFTs each with a limited set of fixed attributes (e.g. a broadleaved evergreen tree, a narrowleaved evergreen tree etc.). New developments allow a much wider diversity of plant attributes within major growth forms, include regeneration modules, and can model trait evolution (Scheiter et al. 2013). DGVMs have their limitations. In some cases this is because of a lack of predictive understanding of the physiology, such as the influence of nutrients other than nitrogen. Simulation of vegetation dynamics impacted by herbivory has also proved challenging (Pachzelt et al. 2015), as has the simulation of global fires, including the very frequent fires characteristic of seasonally humid savannas (Hantson et al. 2016).

Validation of DGVMs was initially based on ecosystem properties such as net primary production. However, at least in principle, DGVMs could also be used to predict major vegetation types from simulated dominant growth forms. There were significant hurdles in doing so, not least the availability of global vegetation maps which could be interpreted in the same terms as the model output. The first generation of DGVMs was also constrained by the small number of growth forms that could be simulated. For example, shrubs and succulent growth forms were not simulated. Nevertheless, DGVM outputs provided the opportunity to formally test what ecosystems should look like if

primarily determined by climate. This is particularly useful in EUCZs to help discriminate between climate regions warm enough and wet enough to grow closed forests from those where trees are unable to grow because of low productivity.

2.6.2 Southern African case study

Southern Africa is an interesting region to test the importance of climate in controlling major vegetation patterns. The main constraint on plant growth is precipitation rather than temperature. The western half of southern Africa is arid while the mesic eastern half is a mosaic of grasslands, savannas, and forests in areas with wet summers and dry winters. In the south-western corner of the region, winter rainfall climates prevail, and the dominant vegetation is shrubland (Chapter 1). Fynbos, a species-rich heathland, is the dominant land cover in higher rainfall regions. Fynbos is considered a Mediterranean type shrubland but large areas receive much higher precipitation than other Mediterranean regions (up to 3000 mm per annum versus 1000 mm for Californian chaparral). Within fynbos landscapes there are also small patches of closed evergreen forest structurally very similar to the forests of the summer rainfall regions (Figure 2.4). All the higher rainfall regions fall into Whittaker's Ecosystem Uncertain Climate Zone and southern Africa exemplifies that zone in the presence of closed forests and open ecosystems in the same landscapes.

The results of a DGVM simulation of southern African vegetation are shown in Figure 2.7 (Plate 4; Bond et al. 2003a). The whole of the eastern part of the region and wetter parts of the winter rainfall regions in the South West are predicted to have high woody biomass consistent with a closed forest. So neither the shrublands of winter rainfall regions nor grasslands and savannas of summer rainfall regions are at equilibrium with climate. Vegetation fires are common in southern Africa and have long been considered to be a major factor influencing vegetation patterns (Philips 1930; Wicht 1948). Could it be that the mismatch was not the fault of the physiological modelling but the assumption that climate alone accounts for the distribution of the vegetation? A fire module was included in the DGVM which

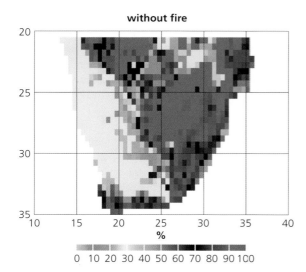

without fire

%

0 10 20 30 40 50 60 70 80 90 100

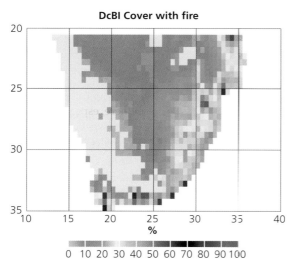

DcBl Cover with fire

%

0 10 20 30 40 50 60 70 80 90 100

Figure 2.7 Tree cover in southern Africa simulated by a DGVM. The figure above indicates predicted tree cover based on the climate potential for tree growth. The figure below includes a fire module. The actual vegetation is grassland and savanna in the summer rainfall climates of the North East and fynbos shrublands in the winter rainfall climates of the South and South West. Small patches of forest occur in both climate zones consistent with vegetation simulations w thout fire. From Bond et al. (2003a). (See Plate 4.)

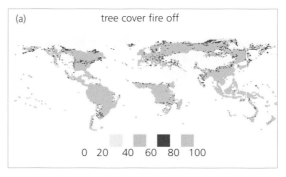

(a)   tree cover fire off

0   20   40   60   80   100

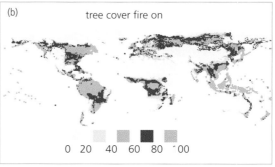

(b)   tree cover fire on

0   20   40   60   80   ʼ00

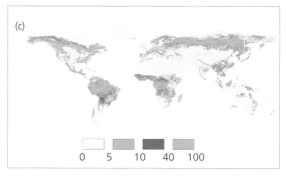

(c)

0   5   10   40   100

Figure 2.8 Global tree cover (%) (a) simulated with 'fire off' (b) simulated with 'fire on' (c) observed tree cover derived from satellite imagery in 2000. Simulated cover values are median tree cover for twentieth century simulations. Observed tree cover classes are: 40–100%, closed forest with no grass understorey; 10–40% more closed forms of savanna and other types of 'forest'; 5–10% scattered trees. The map does not discriminate between natural forests and plantations. Figure from Bond et al. (2005), *New Phytologist*, with permission. (See Plate 5.)

simulated reasonable fire return times for the region. Simulations with the fire module 'on' were a much closer approximation to the actual vegetation than simulations with fire 'off' (Figure 2.7). At least in this

region, dominant plant cover cannot be predicted from physiological responses to climate. Instead, the simulations support the view that fire is a major factor determining what biomes occur where.

2.6.3 Simulating world vegetation

When extended to a global scale, DGVM simulations of the world as determined by climate, with no fire, generated at least double the forest area estimated by FAO (Figure 2.8; Plate 5; Bond et al. 2005). DGVM predictions of above-ground stem biomass with no fire were validated by comparison with data for forest biomass and planted plantations, and changes in woody biomass in experiments with long-term fire exclusion.

DGVMs are a major new tool for testing long-standing hypotheses on the relationships between climate and vegetation (Scheiter et al. 2012; Moncrieff et al. 2016). Because they are mechanistically based, these vegetation models allow us to separate correlation from cause at large spatial scales. As one example, remotely sensed tree cover, as discussed above, was identified by Staver et al (2011a) as deterministically savanna, deterministically forest, or a mixture of the two. However, this classification is not based on physiological constraints on tree growth but on the patterns visible at the particular scale of the satellite products. DGVM modelling indicates that vast areas of the 'deterministic savannas' are warm enough and wet enough to support forest in Africa. Indeed, foresters have unwittingly tested these predictions by planting conifers and eucalypts, among other tree species, very successfully, while small indigenous forest patches are ubiquitous in these landscapes as indicated by ground-based vegetation maps (Lehmann et al. 2011).

2.7 Explanations for uncertain ecosystems

The common assumption that one climate = one vegetation type is clearly wrong for very large areas of the world. The evidence comes from empirical field studies (Whittaker 1975; Bowman 2000; Bond 2005; Lehmann et al. 2011; Dantas et al. 2016; Erdös et al. 2018), satellite imagery of tree cover at continental scales (Bucini & Hanan 2007, Sankaran et al. 2008; Staver et al 2011a,b; Hirota et al 2011; Murphy & Bowman 2012; Xu et al. 2016), and tests of predictions of tree cover based on simulation models based on physiological processes (Woodward et al. 2004; Bond et al. 2003a,2005; Scheiter & Higgins 2009).

2.7.1 Are uncertain ecosystems merely the product of deforestation?

Of course, the existence of open vegetation, especially grasslands and savannas, in climates warm enough and wet enough for forests has long been recognized and the causes of the mismatch with climate have generated heated debate among generations of ecologists. However, the debates have generally been restricted to a regional context. The global scale of the mismatch between vegetation and climate has only become apparent since the 2000s.

Within regions, deforestation by people has been by far the most common explanation for the existence of low-biomass, open vegetation in landscapes that also support closed forests. Anthropogenic deforestation may occur either directly by felling or indirectly by burning. The history of deforestation has been well documented in parts of Europe and North America (Williams 2003). The story has been assumed to be the same elsewhere, especially in the tropics, where large areas of open vegetation occur. The immediate problem with the deforestation narrative is that it fails to explain the high diversity of plants and animals restricted to open habitats. Open vegetation includes some of the 'hottest' biodiversity hotspots in the world. They include cerrado (Brazilian savanna) (Ratter et al. 1997), the species-rich montane grasslands of Brazil, Africa, and Asia (Bond & Parr 2010; Overbeck et al. 2007; Fernandes 2016), and the grasslands and woodlands of the North American Coastal Plain (Noss et al. 2015). The open heathlands of the Southwest Cape of Africa (Goldblatt & Manning 2002; Linder 2003) and south-western Australia (Hopper & Gioia 2004) are the richest temperate zone floras in the world (Cowling et al. 1996). It is not possible for all these species to have evolved in the few centuries or millennia of human deforestation. Both fossil and phylogenetic evidence indicate that the floras of these open systems are millions of years old (Chapter 5; Linder 2005; Crisp et al. 2009; Simon et al. 2009; Maurin et al. 2014; Lamont & He 2012; Rundel et al. 2016; Carpenter et al. 2017). What is more, open ecosystems were even more extensive in tropical zones in the last glacial when human populations were either tiny or did not exist in the case of the Americas (Dupont et al. 2000; Mayle et al. 2004). There can be no doubt that secondary grasslands created by deforestation are widespread and that they are increasing under the pressure of expanding human populations. However, it is clear that open vegetation in climates suitable for forest occupy extensive areas of the world and did so long before human influence. Natural open ecosystems have been ignored for far too long, with serious consequences for their management and conservation. In the following chapter, I will explore the concepts emerging to explain their existence and some implications for ecological theory.

CHAPTER 3

Uncertain ecosystems: the conceptual framework

3.1 Introduction

Stephen Carroll, in his highly readable account of the 'Serengeti rules' (Carroll 2016), notes the importance of detecting and understanding regulation of key processes. He discusses regulation, and the consequences of its breakdown, in the human body but also in ecosystems, and especially in the savannas of the Serengeti. The Serengeti is particularly apt because of the puzzling switch from a wildebeest population hovering around 50 000 animals in the 1960s rocketing to the massive herd of 1.5 million by the 1980s. Clever detective work found the cause of the low numbers to be the rinderpest disease and the massive population spurt to be a side-effect of controlling the disease in cattle (Sinclair 1979). Without understanding the regulatory processes, ecologists will be floundering in trying to forecast the future of any ecosystem and managing its trajectories. It is perversely comforting to know that medical researchers were equally in the dark over common medical problems before they understood regulation of key physiological processes. So, what do we know about the regulation of trees, especially when grouped into forests? And how well do we understand the absence of trees in climates apparently suitable for them?

Finding answers to what regulates tree cover touches on some of the key issues in ecology and has repercussions for ecological theory, environmental policy, and management. The key concepts are introduced in this chapter and provide the context for subsequent chapters in the book.

3.2 The carrying capacity for trees

The key starting point is to establish whether an open ecosystem has the growth potential to form a closed forest. To do this we need an estimate of 'carrying capacity' of trees in a given environment (Figure 3.1). For animals, carrying capacity is measured as the population size at which the population remains stable over time because limited resources constrain further growth. For trees, population size is commonly measured as per cent cover, basal area of stems, or above-ground biomass. Where cover is used for tree importance instead of biomass or basal area, carrying capacity will asymptote since it cannot exceed 100 per cent. The carrying capacity for trees can be measured empirically as the maximum stem biomass (or cover, basal area) for a given region and extrapolated across the region (e.g. Greve et al. 2013). Where no native closed canopy vegetation can be found, forestry plantations can be a useful indicator of potential carrying capacity for trees (Figure 1.2). The development of DGVMs (Chapter 2) has been particularly useful in estimating potential woody biomass over large scales using physiological principles. DGVMs escape the problem of confounding correlation with cause in climate/vegetation patterns and can also be applied globally to diverse climate regions. DGVMs have been important in exposing the extent of the mismatch between climate and potential vegetation (Chapter 2).

Open Ecosystems: ecology and evolution beyond the forest edge. William J. Bond, Oxford University Press (2019). © William J. Bond 2019.
DOI: 10.1093/oso/9780198812456.001.0001

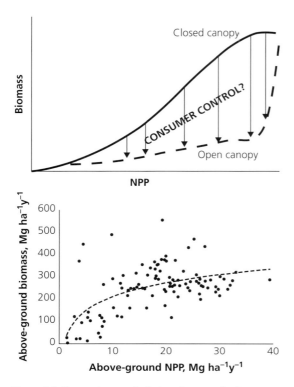

Figure 3.1 The carrying capacity for trees is expected to increase along a productivity gradient. Top: The difference between potential tree biomass (closed-canopy vegetation) and actual biomass (open-canopy vegetation) requires explanation. The difference (vertical arrows) is often due to consumer control by herbivory or fire. Bottom: Actual forest biomass along an NPP gradient across all biomes (from Keeling & Phillips (2007), Figure 4 b, dashed line is their asymptotic regression fitted to the data).

3.3 Explanations for open, low biomass ecosystems: Bottom-up factors

3.3.1 Climate extremes

Simple climate correlations, such as Whittaker's temperature/precipitation plane, use annual averages to predict biomes. But climate extremes, such as severe drought or frosts, may be more important than average conditions in structuring ecosystems. This might especially be so if extreme events, though rare, are sufficiently frequent to occur within the lifespan of dominant trees. Fensham and Holman (1999) have provided an example from Queensland

in north-eastern Australia where *Eucalyptus melanophloia* dominates savanna woodlands. Periodic extreme droughts have been observed which kill mature trees, re-setting the system which subsequently recovers from surviving juveniles. Fensham et al. (2009) have argued that episodic severe droughts account for low tree cover in Queensland and in savannas elsewhere in the world; it is heavy mortality of mature trees during episodic droughts that maintains the open structure of savannas rather than fires burning the grassy understorey. However, extensive drought deaths in savannas elsewhere are seldom reported, and nor are they a feature of the extensive eucalypt savannas of northern Australia. Cyclones cut swathes through mature trees in these northern Australian savannas (Cook & Goyens 2008) and also occur in the pine savannas of the southern USA. Simulation models show they do contribute to maintaining open pine savannas in conjunction with fire which is the major process (Beckage & Ellingwood, 2008). Cyclone damage in tropical forests has long been investigated but has not yet been credited with creating open-canopy from closed-canopy vegetation.

Episodic frost has also been explored as a factor limiting tree populations in various open vegetation types, including chaparral (Mooney 1977) and African savannas (Holdo 2007; Whitecross et al. 2012). Though frost may certainly reduce trees locally in these systems, a transplant experiment across an altitudinal gradient found that frost could not explain the general tree deficit in open grassy ecosystems except at the highest elevation grasslands (Figure 3.2; Wakeling et al. 2012). *Acacia karroo* (=*Vachellia karroo*), also showed evidence of local adaptation with plants grown from its frosty upland locations not suffering from frost. These grasslands form mosaics with closed-canopy forests, further indicating that frost death (along with other 'meteorological disasters') does not account for the general absence of trees in upland grasslands (Wakeling et al. 2012).

3.3.2 Hostile soils

Soils may exclude trees where they are too saline for tree growth or where cracking clays (vertisols) tear

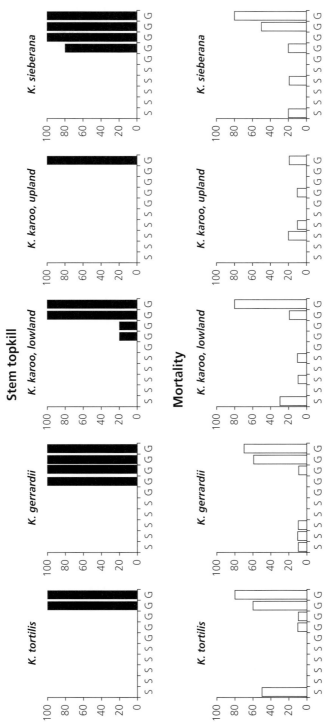

Figure 3.2 Frost effects on seedlings of four savanna tree species along an elevation gradient. The degree of topkill and percentage mortality were recorded after winter for each species at each of 10 sites. None of the savanna sites, and only the top four grasslands sites, experienced frost. There were 10 individuals per species per site. Top: proportion of plants with >75% topkill after frost. Bottom: proportion of plants dying after frost. The first five sites were savannas, S, and the next five sites montane grasslands, G. Data from Wakeling et al. (2012).

roots inhibiting tree growth. Anoxic conditions in the rooting layer caused by waterlogging also exclude trees. These physical constraints on tree growth are common and widespread but seldom extend over large geographic regions. Rare exceptions large enough to be mapped at global scale include the seasonally flooded Pantanal of Brazil and the Mitchell grass plains of north-eastern Australia with extensive black cracking clays. Chemical (nutrient) constraints on forests have been a popular explanation for open ecosystems since forests generally occur on nutrient-rich soils. The general problem with the hypothesis is teasing out whether nutrient differences are a cause or consequence of vegetation type. Until recently, mechanistic explanations for why forests, with denser, shadier trees, should require higher nutrients, have been lacking. Soil-based hypotheses for open-canopy vegetation are discussed further in Chapter 6.

3.3.3 Physical disturbance

All forests are exposed to disturbances whether at the scale of a single tree fall creating a light gap or a system-wide catastrophe such as major cyclone damage, floods, or volcanic eruption. Are open-canopy ecosystems merely 'early successional stages' recovering from catastrophic disturbance? The stability (age) of the open system can be a useful guide. Studies of primary succession, even in hostile and unproductive climates such as Glacier Bay (Chapin et al. 1994), indicate that forests develop within 200–300 years. If a system has remained in an open-canopy state for centuries to millennia, then calling it an 'early successional' stage is nonsense. To accommodate vegetation states that were apparently stable but not a 'true' climatic climax, Clements (1936) developed a complex taxonomy of different kinds of climax. Different prefixes indicated the most typical obstacle thought to be preventing the system from reaching its climate potential, such as 'fire subclimax'. The invention of these new terms helped prolong the use of successional concepts in circumstances where they were inappropriate instead of challenging ecologists to develop more appropriate concepts for 'early successional' stages that stay 'early successional' for millennia.

3.4 Anthropogenic disturbance and open, low biomass ecosystems

Deforestation by felling and burning is by far the most popular explanation for anomalous open systems where the climate can support forests. Where the cause of 'deforestation' is not obvious, deforestation is usually attributed to subsistence farmers or hunter-gatherers and their damaging use of fire to manage their environments. Williams (2003) summarized the evidence for deforestation in both temperate and tropical regions, emphasizing the destructive power of fire. Major human influence on forests is comparatively recent, peaking only in the last 10 000 years. Human use of fire is much older than the Holocene, but the environmental impacts are uncertain (Archibald et al. 2012). Deforestation is occurring widely today and the rate of deforestation is increasing, especially in countries with burgeoning human populations. In Madagascar, for example, the human population grew from ~2 million in the early 1900s to >27 million in 2017. Pressure on the remaining forests has grown in synchrony and deforestation is now widespread. But more than half of Madagascar is covered by tropical grasslands and arguing that deforestation on the present scale produced this vast area of grassland by the early twentieth century is another question entirely (see Chapter 4).

To test a deforestation origin for open systems, we need to test the antiquity of open habitat using historical and prehistorical information. Diverse sources of information are becoming available in addition to palynological studies (e.g. Aleman et al. 2018). Fairhead and Leach (1995) used historical documents to explore the origins of a 'pristine' forest in West Africa declared a World Heritage site for its pristine status. These showed that, far from being an ancient forest, the area had been a savanna in the nineteenth century! Carbon isotopes are a powerful tool for indicating past vegetation in tropical regions. Soil organic matter preserves the isotopic signal of the source material and clearly distinguishes tropical (C_4) grasses from woody vegetation (e.g. Cerling et al. 2011). For more detail on grass and tree composition, phytoliths can be used. These are silica bodies in the epidermis of plants which vary greatly in shape, providing useful taxonomic information (Figure 3.3).

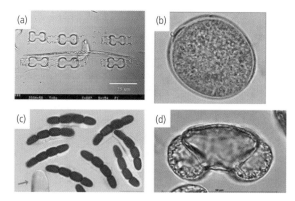

Figure 3.3 The stability of open- and closed-canopy ecosystems can be tested by a variety of paleo-proxies for vegetation type: (a). phytoliths, (b) grass pollen, (c) *Sporormiella*, a coprophilous fungus, (d) tree pollen (*Podocarpus*).

An indirect test of anthropogenic or ancient origins of open systems is to explore the diversity and endemism of open habitat biotas. 'Old growth' tropical grasslands, with no known record of cultivation, differ from secondary grasslands in having much richer plant diversity containing far greater numbers of long-lived perennial forbs with underground storage organs (Chapter 4; Veldman et al. 2015a). Plants and animals that are endemic to open habitats are common in the 'semi-natural' grasslands of Europe, suggesting their habitats have been misclassified as of anthropogenic origin. Cape fynbos shrublands, containing the world's richest temperate flora, are ancient (Chapter 1), yet occur side-by-side with tall closed forests. The implications of an endemic, open-habitat flora and fauna, quite different from forest floras, has yet to be widely appreciated. If you observe a species-rich assemblage restricted to 'early successional', open vegetation, then it is not a leftover from the forest but, more likely, a remnant of an open, ancient ecosystem (Chapter 4).

3.5 Consumers and open, low biomass ecosystems

3.5.1 The green world hypothesis

Bottom-up control, firstly by climate and secondly by soils, has been by far the most widely accepted explanation for vegetation distribution for centuries.

This world view was shaken by Hairston et al. (1960), in their brief (5 page) paper in *American Naturalist*. They asked the question 'why is the world green'? The question is not about the spectral properties of leaves but about regulation of the amount and kind of plant matter eaten by herbivores. Why don't herbivores eat so much vegetation that they alter the amount of plants, changing the structure and composition of plant communities, and thereby changing their shade of green? Their tentative answer was that herbivores are controlled by predators (or pathogens) and seldom reach densities sufficiently high to regulate vegetation. It is unsettling, for a vegetation ecologist, to consider that the mix of trees and grass in an African savanna depends on the number of lions eating the herbivores that eat the plants. Nevertheless, the jump from 50 000 wildebeest to ~1.5 million in the Serengeti was attributed to control of rinderpest, a disease transmitted by cattle (Sinclair 1979). The wildebeest population explosion had numerous cascading effects on the ecosystem, including reduction of fire because of the reduction of grassy fuels by vast herds of grazers (Figure 3.4; Holdo et al. 2009).

A tricky problem with the green world hypothesis is that if predators (or pathogens) did regulate herbivore pressure to such an extent that their effects were negligible, then you would not be able to tell the difference between vegetation controlled from the top down, or from the bottom up, by climate and soils! Trophic control would only be revealed by eliminating predators. The reintroduction of wolves in North America, such as in Yellowstone, is an example of trophic control in action. Changes in the behaviour of elk in response to the return of wolves led to saplings escaping browse damage which then transformed a meadow into a woodland (Ripple & Beschta 2004).

3.5.2 Consumers that escape predation

Not all herbivores are vulnerable to population regulation by predators and these herbivores should be particularly influential in shaping ecosystems. Owen-Smith (1988) argued that megaherbivores, animals >1000 kg, escape predator regulation of their populations by being too large and fierce to provide regular prey. Examples include elephants,

(a)

(b)

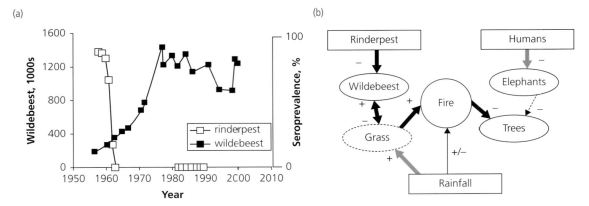

Figure 3.4 Wildebeest numbers have cascading effects on the Serengeti ecosystem. (a): Wildebeest populations (solid dots) were released by control of rinderpest (seroprevalence, open squares). (b): Wildebeest reduce grass biomass which, depending on rainfall, reduces fire activity, releasing tree populations which are also impacted by elephants. From Holdo et al. (2009).

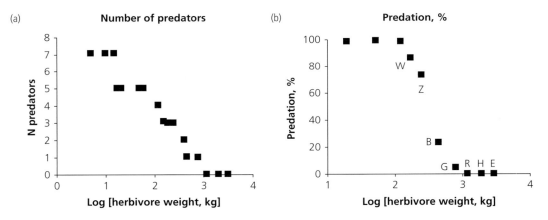

Figure 3.5 Predation varies with body size of the prey in the Serengeti. (a): The number of species of predators preying on herbivores decreases with increasing prey body size. No predators prey on herbivores larger than 1000 kg. (b): The proportion of adult mortality accounted for by predation in non-migratory ungulates. Above about 150 kg, predator limitation switches to food limitation. Symbols refer to species: W, wildebeest; Z, Zebra; B, buffalo; G, giraffe; R, black rhino; H, hippo; E, African savanna elephant. From Sinclair et al. (2003).

rhinoceros, and hippos. These very large animals, he argued, would grow in population size until limited by resources at which stage they would be particularly influential in shaping the ecosystem. Large mammals that live in herds also gain protection for predation. In African savannas, buffalo are fearsome prey even for lions, the largest predators (Sinclair et al. 2003; Figure 3.5). Migratory mammals can also escape predator control if their

predators cannot follow the herd while caring for their young. In all these instances, we might expect mammal consumers to reach densities where they have the potential to regulate tree numbers and maintain open ecosystems. Similarly, extirpation of predators should lead to herbivore populations exploding to levels where they maintain open systems. The nature of herbivore impacts is discussed further in Chapter 7.

3.5.3 Fire as a generalist herbivore

Until recently, the trophic ecology literature did not mention fire, our modern world's most widespread and influential consumer. Fire can be viewed as a generalist herbivore with effects analogous to mammal browsing and grazing (Bond & Keeley 2005). Fire invites analogies to herbivory not least because, unlike other physical disturbances such as floods, landslides, and cyclones, fire consumes complex organic compounds and converts them to combustion by-products. Fire has no 'predator control' and, in that sense, escapes the constraints of the green world hypothesis. The preferred 'diet' of fire differs from herbivores in having no nitrogen requirement and in readily consuming plant material high in cellulose and lignin. However, fire also has distinct 'dietary preferences' and is strongly influenced by the moisture content and spatial distribution of plant material. Consequently, fires burn more readily through some vegetation types than others. The frequency of fire, its intensity, seasonality, and spatial extent vary greatly in different ecosystems and climate zones across the world (Archibald et al. 2013). These patterns of burning are called fire regimes, or pyromes at large spatial scales, and strongly influence the nature of the plants that occur within a given regime.

Although there are analogies with herbivory, of course there are also differences. Whereas chronic browsing may be required to prevent trees growing beyond browse height, the action of fire is episodic. Fires depend on a continuous fuel bed (i.e. continuous plant cover) and on the flammability of the dominant growth forms. Fires are common where decomposition is slow due to leaf properties and microclimate. Fire spread also varies with microclimate. In EUCZ regions, fires in open ecosystems usually die out after penetrating only a few metres into closed forests. The failure of fires to spread through closed forests is due to the lack of flammable fuel in the understorey and to the forest microclimate. Savanna grasses cannot survive in dense shade while the forest microclimate, especially reduced windspeed, inhibits fire spread (van Wilgen et al. 1990; Hoffmann et al. 2012b; Little et al. 2012).

3.5.4 A multi-coloured world

Bond (2005) suggested expanding the domain of the 'green world' hypothesis to encompass fire and herbivory as major consumers of vegetation that can alter the nature of vegetation, its shade of green, at a global scale. Vegetation can potentially exist in three states: green world where climate is the major control, brown world where vertebrate herbivory is the major regulator of trees, and black world where fire is influential, and the fire regime regulates the mix of growth forms. Box 3.1 elaborates on the multi-coloured world and its implications.

Box 3.1 Green, Black, or Brown Worlds

Climate has long been considered the major control on the vegetation types of the world, determining the availability of moisture and energy for plant growth. Soils modify plant moisture supply and vary in nutrient supply and are secondary determinants influencing local vegetation patterns. In 1960, Hairston, Smith, and Slobodkin ('HSS') made the radical suggestion that herbivores shape vegetation unless regulated by predators or pathogens. They asked a novel question: 'Why is the world green? Why is it not 'brown'?', or at least a different shade of green as herbivores consume plants? Their answer was that the world is as green as it is because herbivores are regulated by carnivores, pests, and pathogens. Without such regulation, herbivore numbers would explode and the world would no longer be 'green'.

HSS emphasized predation as the central process influencing plant biomass by limiting herbivores. However, the real issue is whether consumers can ever reduce plant biomass enough to significantly affect vegetation structure and composition. Critics of the hypothesis argued that the world is green because most plants are inedible, full of indigestible components such as cellulose and lignin (Polis 1999). Or

continued

Box 3.1 *Continued*

there is insufficient nitrogen, essential for building animal protein, to sustain significant herbivore pressure (White 2005).

However, food quality requirements vary with body mass and large mammals can consume low quality diets. Large animals are also difficult and dangerous prey for predators. For these reasons, Owen-Smith (1988) suggested that megaherbivores (animals with body mass >1000 kg) would be particularly influential in shaping vegetation structure. Megaherbivores roamed most of the planet just a few thousand years ago. They would have been prime candidates for significant consumer-control of vegetation (Owen Smith 1988). However, smaller mammals, including goats or even rabbits, exert significant consumer-control in many ecosystems by consuming seeds, seedlings and saplings thereby reducing woody recruitment (Augustine & McNaughton 1998).

Fire has been missing in the ecological literature but can be viewed as a generalist consumer with properties, and effects, like animal consumers (Bond & Keeley 2005). Fire consumes large amounts of plant material in irregular 'bites' of varying intensity and in preferred seasons. Like animal consumers, but unlike other physical disturbances, fire consumes complex organic compounds and converts them to combustion by-products. Fire differs from mammalian herbivory in defoliating vegetation regardless of quality and does not require protein for growth. Fire thrives on a 'diet' of plants with properties making them inedible to mammal herbivores—high cellulose and lignin and low nitrogen content. Indigestible plants are the 'food' that feeds fires.

The HSS question now becomes: Where is the world green? And where is it not green? Is it brown because of heavy herbivory? Or black because of regular fires? Bond (2005) labelled 'brown' world as where intense mammal grazing and browsing control vegetation; 'black' world where fire is a regular and influential process, and 'green' world where climate, and not consumers, limit woody biomass and the plants approach their climate potential (Box Figure 3.1). Transitions between the different coloured worlds can occur. Since fire and large mammal grazers are competing consumers of vegetation, one can ask whether more of the world has become 'black' since extirpation/extinction of the megafauna (Flannery 1994). Or has the replacement of the megafauna by domestic livestock compensated for the loss of an entire class of consumers? Is the poor forage quality and high productivity of, say, humid tropical grassy ecosystems or Mediterranean-type shrublands such that fire is always the dominant consumer in higher rainfall regions? Or could mammals replace fire as major consumer?

The three-coloured world is a useful framework for exploring the world. Removing herbivores should have major cascading consequences in brown world but not in black or green worlds. Suppressing fire should have cascading consequences in black world but not in brown or green worlds. The possibility of finding all three worlds would also be expected to vary depending on climate, soils, and environmental history. Functional traits would be expected to differ in each of the three worlds because of the very different selection pressures.

Box Figure 3.1. A multi-coloured view of the world. Communities in any locality may have elements of all three possible ecosystem states depending on the history, magnitude, and type, of consumer-control. The resource base influences the probability of transitions from one state to another. For example, nutrient poor soils would tend to reduce mammal herbivory, favouring fire and 'black world' species. Redrawn from Bond (2005) Journal of Vegetation Science.

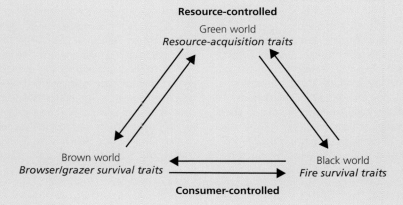

Resource-controlled

Green world
Resource-acquisition traits

Brown world
Browser/grazer survival traits

Black world
Fire survival traits

Consumer-controlled

3.6 Integrating top-down and bottom-up processes

Aquatic ecologists have long accepted that lake properties are controlled by both bottom-up and top-down processes (Carpenter et al. 1985). The rapid response of aquatic systems to experimental manipulations no doubt helped them understand trophic control decades earlier than their terrestrial colleagues. Responses of terrestrial vegetation to experimental manipulations are typically one or two orders of magnitude slower. It takes at least twenty or thirty years of fire suppression to convert a savanna to a forest, and a century or more to convert fynbos shrublands to forest. Thus, the evidence for interactions between top-down and bottom-up controls is primarily explored in ecological models supplemented by experiments, planned and natural.

Kellman (1984) was among the first to provide an integrative hypothesis for the association of tropical forests with nutrient-rich soils and savannas with nutrient-poor soils. He argued that forest trees would grow faster on richer soils reaching fire-proof size earlier than on nutrient-poor soils. Forests would also recover faster on the richer soils if they did burn. On poorer soils, colonization and recovery would be too slow to form fire-proof forests. Hoffmann et al. (2012a) have extended this model based on time taken to reach threshold sizes, one threshold relating to fire resistance, the other to fire suppression (Chapter 7). Trees need to grow to a threshold size before gaining resistance to fire damage. Savanna trees invest more in structures such as thick bark enabling them to reach this threshold faster than forest trees. The second threshold is the size at which a tree shades out understorey grasses thereby resisting the spread of fire. Forest trees tend to cast more shade than savanna trees and may reach this threshold earlier for a given tree size. The rates of growth to both thresholds is a function of site conditions while fire frequency and intensity sets the time required to reach the threshold (Figure 3.6). The model was developed in Brazil but is being extended to tropical and temperate systems elsewhere to help define boundary conditions in forest/non-forest mosaics.

Similar considerations apply where mammal herbivores are the consumers (Chapter 8). Plants have to grow to a threshold size above the reach of browsers, equivalent to the 'resistance threshold'. The 'suppression' threshold has no formal equivalent where mammals are consumers. However, shade-intolerant herbaceous understoreys of ferns and grasses suppress tree regeneration in many temperate woodlands (Chapter 8). Figure 3.6 shows an example of the height threshold for escaping deer browsing in *Pinus sylvestris* and the influence of site conditions on reaching that threshold in upland Britain (Palmer & Truscott 2003). The marked effect of site on the rate, and therefore the probability, of reaching size thresholds are similar to those for the fire-based models of tree recruitment in open ecosystems (Higgins et al. 2000) and trees on the boundary of the forest/savanna mosaic (Hoffmann et al. 2012a).

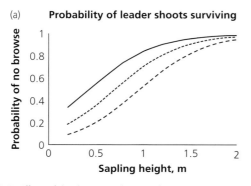

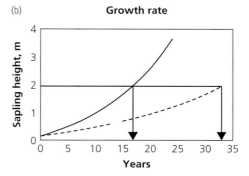

Figure 3.6 Effects of deer browse and site conditions on Scots pine growth. (a) shows probability of leader shoot survival in relation to sapling height: solid line, red deer 5 km^{-2}; dotted 10 km^{-2}; dashed, roe deer 5 km^{-2}. (b) shows growth rate of unbrowsed saplings to 2m escape height: solid line, dry mineral soil, good year (17 y); dashed line, wet, peaty soil, good year (32 y). Data from Palmer & Truscott (2003).

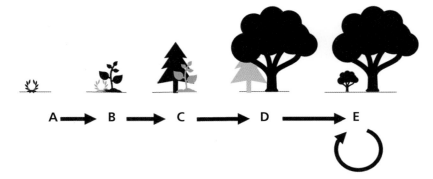

Figure 3.7 Successional change driven by habitat modification by early successional species (facilitation) and competition for light. Competition for light is asymmetric—taller plants access more light than shorter plants. Succession proceeds until trees casting the most shade dominate a forest and whose seedlings can recruit in their shade.

3.7 Open ecosystems and succession mechanisms

Classical successional theory, developed by Clements (1936) and others, emphasized facilitation and competition as the central mechanisms driving the process (Connell & Slatyer 1977). Since competition for light is asymmetric (taller plants gain disproportionately more light than shorter plants) succession should proceed by more shade-tolerant plants successively replacing less shade-tolerant early-successional plants. This logic predicts that, where the climate is suitable, succession will proceed until forests become the dominant natural vegetation (Figure 3.7). Indeed, it is with this logic that ecologists from the forested landscapes of Europe and Eastern North America first viewed the extensive grasslands of the tropics. Deforestation in Europe had produced open vegetation and therefore the most reasonable explanation for extensive open grasslands and shrublands in other parts of the world must also be due to deforestation (see, e.g., Williams 2003 for an exhaustive review).

From an evolutionary perspective, the facilitation pathway of succession is puzzling. Each successional stage alters the environment in ways that promote the next successional stage. Thus, the plants of early successional stages are altering the environment in ways that reduce fitness of their own offspring. Wilson and Agnew (1992) pointed out that it is just as probable, indeed more probable from an evolutionary perspective, that plants alter their environment to promote their own preferred conditions. This

has become known as 'niche construction' in the context of a population (Odling-Smee et al. 1996, 2003) and 'ecosystem engineering' when 'the activity of organisms that create, alter or destroy habitats' modifies resource availability to other species (Jones et al. 1994; Erwin 2008).

If feedbacks to the environment help maintain a species by constructing its preferred conditions, then it is possible that a site with the same physical conditions (climate, geological substrate) could end up supporting very different ecosystems depending on the way in which the plants alter the environment. We might therefore expect that plants of open ecosystems alter the environment in ways that help slow or prevent colonization by closed forest trees. The nature of the feedbacks should be sufficiently strong and general to account for the large global extent of open ecosystems (see Wilson & Agnew, 1992, for a comprehensive review).

3.8 Alternative stable states

Both resources and consumers are nicely integrated in Alternative Stable State theory (ASS) which is increasingly being applied to closed and open ecosystems in the tropics and elsewhere. The theory emerged from theoretical population biology some 50 years ago as a possible outcome of population models which predicted more than one stable equilibrium point (Lewontin 1969, May 1977). It has since attracted a large and growing literature to which the reader is referred for more advanced treatments (Scheffer & Carpenter 2003; Scheffer et al. 2012a; Warman & Moles 2009; Petraitis 2013). As applied to ecosystems, ASS differs from classic resource-centric ideas. Rather than an invariant environmental setting deterministically shaping the vegetation, ASS emphasizes environmental differences caused by biotic feedbacks to the environment. ASS seems particularly appropriate for analysing open versus closed ecosystems (Warman & Moles 2009; Staver et al. 2011a; Hirota et al. 2011; Wood & Bowman 2012). Indeed, recognition of tropical grassy and forested systems as bistable states has dragged ASS from an arcane theory of diminishing relevance in ecology (Petraitis 2013) to a major paradigm for understanding the geography of world vegetation (Staver et al. 2011a,b; Hirota et al. 2011).

Alternative Stable State theory (ASS) has been applied to a wide range of aquatic and terrestrial ecosystems (Scheffer & Carpenter 2003; Petraitis 2013). In such systems, two or more states can occur; such as a lake with one state of dense phytoplankton blooms and the alternative state of clear water and macrophytes. Each state is **stable** in that it returns to the same state after minor perturbations. States are not permanent and can switch to the alternative state, termed a 'regime shift'. Regime shifts can be recognized by a major change in a system variable that defines the alternative states, such as stem biomass in open versus forest vegetation. ASS is also often represented by a 'ball and cup' analogy. The ball represents the community or ecosystem and each cup represents a basin of attraction. Different ecological regimes develop, each with its own basin of attraction maintained by strong feedbacks. The existence of two basins of attraction is called 'bistability'. Each state is resistant to regime shifts and discrete states persist until an ecological threshold is crossed. Shifts from one regime to the other can occur in two ways:

1. When there is a major shock to the system so that the feedbacks that usually maintain the system are overwhelmed causing a sudden shift to the alternative state.
2. There is a gradual change in system drivers. Here the basin of attraction gets shallower until a threshold is crossed and the system disappears and switches to the alternative state.

An example for closed forest versus open, fire-maintained grasslands is shown in Figure 3.8.

Alternative stable states are characterized by hysteresis. This is when the system state is determined by environmental conditions that existed in the past and not only at the moment of observation. Hysteresis emphasizes history since the point at which a system flips is not the same as the point of switch in the reverse direction. For instance, grazing may maintain a savanna, and removing grazers may drive the system to a forest, but the savanna is not recovered by just adding back grazers to the forest.

ASS have several notable features. System history is important, as emphasized by hysteresis. Small changes in environmental drivers can produce large changes in ecology. Predicting system shifts is difficult since the future is not a linear extrapolation of the past. Threshold conditions, 'tipping points', are the lines of fracture where the avalanche of changes begins. The resilience of the system to change can be visualized as the depth of the cup. A deep cup is very resilient, a shallow cup can easily shift to the alternative state. Methods have been developed to attempt to recognize when a system is approaching a regime shift to an alternative state using the changing variance in key ecosystem variables and the slower recovery rate to equilibrium as the system approaches a tipping point.

3.8.1 Testing ASS

Despite theoretical advances, the existence of ASS remains controversial. Petraitis (2013), in an extensive review of ASS theory and tests, was pessimistic as to the utility of the theory, arguing that almost no convincing examples had been demonstrated despite decades of effort. He described it as 'one of most interesting and vexing problems in ecology'. If they exist, ASS are important to understand since ecosystems that appear persistent are susceptible to catastrophic change (Scheffer & Carpenter 2003). The unpredictable nature of shifts and the difficulty of reversing changes easily (hysteresis) can cause serious environmental problems. But tests of ASS are fraught with difficulties and semantic issues starting with defining the variables that determine the alternative stable states. I have emphasized shade as the key distinction between open and closed vegetation states (see also Ratnam et al. 2011), which is often, but not always, associated with low versus high stem biomass (Pausas and Bond in press).

(a) Savanna fires do not burn beyond the forest margin.

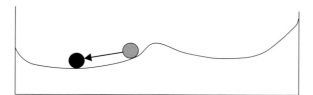

(a) Ordinary savanna dynamics. Tree cover fluctuates from low to high but never shades out grasses

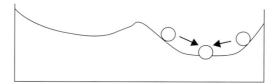

(b) Fire storm penetrates deep into the forest.

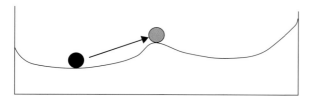

(b) Tree cover increases, grasses still present. Harder to burn with less grass fuel. A warning sign of approaching tipping point.

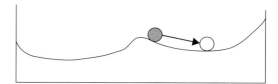

(b) Grasses invade post-burn, fuel follow-up fires killing resprouting trees. System switches to savanna state

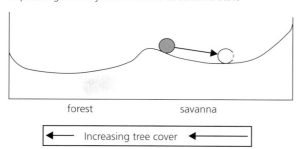

(c) Tree cover and composition changes. Grasses shaded out, fires suppressed. System switches to fire-resistant forest.

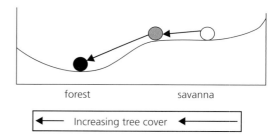

Figure 3.8 Two kinds of regime shifts with forest/savanna examples. Left: Noise-induced tipping where a brief, large perturbation, a severe fire penetrating beyond the forest margin, opens up the system to light-demanding flammable grasses causing a system shift to savanna. Right: bifurcation type tipping where gradual changes in external conditions affect the resilience of the savanna state. Here, closed woody cover increases, gradually shading out grasses, until fires can no longer spread through the landscape and it switches to closed canopy forest. See Lenton (2013).

3.8.2 Alternative hypotheses

Tests of ASS revolve around disproving alternative hypotheses at least as much as testing ASS predictions. Competing ideas for the presence of open and closed vegetation in the same landscape are, first, that the different states represent different successional states. Thus, grasslands and shrublands have long been described as successional to forest. Only if the open system is persistent (stable) can it be considered an ASS. The second major alternative hypothesis is that the two states exist because of environmental conditions external to those created by each state. This has long been the view of advocates of bottom-up explanations for the distribution of forests in otherwise open vegetation. The forests, it is argued, require different environmental conditions—a moister microclimate or a particular soil or particular nutrient concentrations—so that forests will only occur where these conditions are satisfied. Opponents argue that forests have wide environmental tolerance and, if restricted to particular soils and sites, this merely reflects local variation in the disturbance regime and/or its interaction with growing conditions (Bowman 2000; Kellman 1984). Experimental demonstration of ASS in terrestrial systems must thus demonstrate a) that both states are stable and not transitory and b) that each state can replace the alternative state, thereby indicating that environmental conditions can support both states. Given that a forest tree may take twenty or thirty years to mature, the high bar for demonstrating ASS in terrestrial systems has seldom, if ever, been crossed according to Petraitis (2013)!

3.8.3 Experiments

Nevertheless, long-term experiments manipulating fire and large mammal herbivory do exist (Figure 3.9; Plate 6). Many of these have been maintained for

decades and provide key insights into the potential for forest development from open vegetation (Louppe et al. 1995; Laris & Wardell 2006; Bond et al. 2005; Beschta & Ripple 2009; Tanentzap et al. 2011; Veenendaal et al. 2018) (see Chapters 7 and 8). Few of these field experiments would meet the rigorous criteria of replication and data collection required by modern experimental ecologists. But spare a thought for the tasks involved. Just maintaining fire suppression treatments or repairing herbivore exclusion fences for multiple decades requires considerable dedication over periods often exceeding the working life of the field technicians and researchers who set them up in the first place. Consumer manipulation experiments were set up

Figure 3.9 Fire is very influential in shaping ecosystems as revealed by long-term experiments. Top: Longleaf pine savannas, Florida, (a) burned annually for 40+ years, (b) with fires suppressed over that period. Bottom: Savannas in Kruger National Park, (c) arid savanna with, left, >40 years fire suppression, right, burned every 2 years, (d) mesic savanna with, left, burned annually for >40 years, right, fire suppressed. The mesic savanna is switching to a closed thicket. Florida pictures, Winston Trollope; Kruger, W. Bond. (See Plate 6.)

by foresters, conservation managers, and rangeland ecologists in diverse parts of the world. Only a small fraction has been written up in the literature. They are a largely untapped resource that still requires data description, collation, and publication (but see Bond et al. 2003a; Bond et al. 2005; Veenendaal et al. 2018).

3.8.4 Stability

Petraitis (2013) noted that demonstrating that alternative states are stable is among the most difficult tests of ASS. However, at least for C_4 grassy systems, the criterion of stability of the open ('early successional') state can easily be tested using paleo proxies, especially carbon isotope analysis of soil organic matter. Deeper soil layers retain older soil carbon which can be identified as sourced from C_4 grasses and dated as to origin. For example, one study of subsoils in an African forest revealed grass carbon at depth indicating that the landscape had

supported savannas in the past but with a regime shift to forest some two thousand years BP (Gillson 2015). It is these unequivocal indicators of stability and antiquity of ancient grasslands, shrublands, and savannas which have made nonsense of classifying them as early successional stages to forest. The stability of these radically different ecosystems requires a different conceptual framework and ASS would seem to fit the bill (Chapter 8, Figure 8.15).

3.8.5 Patterns from space

Yet ASS theory has remained controversial until very recently because of the stringent requirements for testing the existence of the alternative states. What was missing from the ecological toolkit of the 1980s and early 1990s was easily available satellite data. New global remote sensing data have revived interest in ASS to the point where it is claimed to be a major explanatory framework for vegetation patterns across large areas of the world. Remote sensing

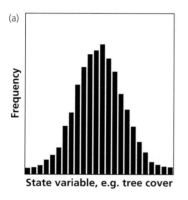
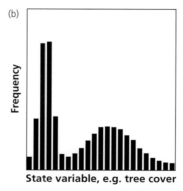
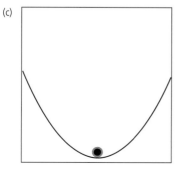
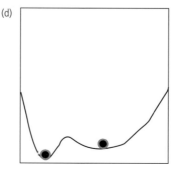

Figure 3.10. Frequency histograms of a state variable such as tree cover (a,b) and inferred stability landscapes below (c,d) with the balls indicating the location of the stable attractors. From Hanan et al. (2014).

studies have revealed that tree cover, or basal area, is distinctly multi-modal: high in forests, intermediate in open savannas and woodlands, or very low as in open grasslands or shrublands. Intermediate cover values are very rare, consistent with ASS (Scheffer & Carpenter 2003). Each mode is viewed as a basin of attraction with intermediate values assumed to be unstable and ephemeral (Figure 2.5). The first analyses interpreting large-scale patterns of woody cover as indicative of bistability appeared as recently as 2011 (Staver et al. 2011a; Hirota et al. 2011). Both were in savannas. Questions on the validity of the analytical methods for detecting multimodal cover from the satellite imagery (e.g. Hanan et al. 2014) have been by-passed by subsequent analyses of tree basal area from plots across Africa demonstrating very similar patterns (Dantas et al. 2016) and by Lidar-based patterns in tree height (Xu et al. 2016). The availability of large climate and soils data sets has also allowed interpretation of the external environmental contributors to the tree patterns at a scale undreamed of by experimental ecologists. Forests and open ecosystems overlap extensively in their climate and soil types consistent with ASS theory (e.g. Dantas et al. 2016).

3.9 Implications of ASS theory

Despite controversy over the existence of ASS and the difficulties of testing the hypothesis, ASS theory is stimulating some radical re-thinking as ecologists digest the implications. Some of these are listed below and compared with older ideas on ecosystem dynamics where appropriate.

1. Disturbances (consumers) are well-integrated into the system in contrast to classic succession theory where disturbance is extrinsic and delays, or reverses succession to the climax.
2. Disturbance regimes not only influence biome distribution but also select for functional traits. The ASS framework is highly suited to explore suites of contrasting traits, for example in black versus green or brown world (Box 3.1). Analysis of these functional differences, in turn, can reveal key processes maintaining alternative states (e.g. Pausas et al. 2004; Charles-Dominique et al. 2015b, 2018; Staver et al. 2012).

3. ASS points to factors other than climate as potential causes of major vegetation change. For example, Bergengren et al. (2011) analysed ecological stability in response to climate change using growth forms that they assumed to be selected by different climates. Their results indicated tropical systems, including African savannas, were the most stable terrestrial region on Earth. Higgins and Scheiter (2012) in contrast, simulated vegetation change in response to climate, fire, and CO_2. Their simulations predicted major shifts from savannas to a wooded state in much of Africa. The region is not stable at all and has the potential for major increase in woody cover with potentially catastrophic consequences for open grassy ecosystems (Parr et al. 2014; Bond 2016).
4. ASS has implications for predicting species responses to climate change. Species distribution models assume individualistic responses of species to climate change independent of other species. But if the species is restricted to a particular biome state, then its responses to climate change will depend on the whole system responding. A forest specialist depends on forest, a grassland specialist on grasslands. Species may behave individualistically within a major biome but spreading to a different biome will only be possible for a small subset of pioneers. The old debate of individualistic versus community unit organization of species clearly needs re-visiting in the context of ASS (e.g. Keddy 2007).
5. ASS emphasizes feedbacks of a system to the environment. So, niche construction or ecosystem engineering become central issues. Plants are not passive responders to fire and herbivory—they can strongly influence fire and herbivore regimes. Fire regimes vary in flammability depending on plant properties. Herbivore pressure varies depending on the quantity and quality of forage and habitat suitability for predator avoidance. Soils are not fixed extrinsic determinants of plant growth. Soils are also products of the plants that grow on them. For years, nutritional research has emphasized nutrient constraints on plant growth (see e.g. Tilman 1982; Craine 2009). ASS forces us to explore plant feedbacks to the soil—how plants influence the nutrients.

6. ASS emphasizes the role of organisms as active participants in changing environments of the past, and not merely responders to changing climates. The appearance of new organisms with new ways of influencing the environment may affect ecosystem changes as profoundly as those initiated by major physical changes in the deep past (Chapter 5).
7. In the present, ASS implies active management of consumers (fire, herbivory) to change or maintain system states. Trophic cascades, leading to massive structural and functional changes, may be triggered by seemingly small and unlikely perturbations. The direct and indirect effects of addition or removal of predators on mammal consumers and they in turn on tree populations provide examples (Beschta & Ripple 2009; Chapter 8).
8. ASS provides clear motivation for conservation of open ecosystems and the processes that maintain them whereas classic succession theory diminishes their value by defining them as 'early successional' in contrast to the climax community.
9. In the ASS framework, history matters. In classic succession theory, systems converge to similar end points regardless of initial conditions. There are several examples of open ecosystems interpreted as continuing legacies of low tree states from the last glacial period when large mammal herbivores were abundant and environmental conditions were hostile to tree growth (Weigl & Knowles 2014; Moncrieff et al. 2014b).

3.10 Summary

This chapter has considered several traditional explanations for open ecosystems (climate, soils) and some very unconventional ones such as large-scale consumer control of world vegetation and the widespread occurrence of alternative stable states. Recognition of open ecosystems as an anomaly to climate explanations of the last two hundred years is the starting point. If they are only products of deforestation then we can shake our heads in sorrow for what is lost. If they are ancient systems,

persisting for millennia by creating their own preferred environmental conditions, then we lift our heads in awe at the ways in which life is organized as a collective in diverse parallel worlds.

The nature of the problem is such that there will not be a single explanation in any region for the existence of open vegetation. Some will be the legacy of deforestation, some will be the remnants of an ancient system and it becomes necessary to identify which is which. Some soils are hostile to tree growth so that fire suppression will leave the system unchanged whereas on different soils the entire open ecosystem might disappear. We have yet to catalogue the extent of consumer versus edaphic control of open ecosystems and to classify which soils are physically incapable of supporting forests and which can be modified in their properties by feedbacks to the ecosystem.

The idea of a multi-coloured world, black, green, and brown (Box 3.1), invites analyses of their spatial extent and of functional and compositional properties characteristic of different coloured worlds. Exploration of the functional traits required to thrive in different fire regimes is currently an active area of research. Traits that allow trees to thrive in brown world are also beginning to be revealed (Chapter 4). Traits important for consumer-driven worlds are not the same as functional traits important for resource acquisition (such as the leaf economic spectrum) but are as profound in filtering which species occur in which coloured world.

Viewing open versus closed ecosystems as alternative stable states has led to an explosion of studies of satellite-derived patterns of tree densities with novel theoretical models to explain the patterns. While originally applied in the tropics, research on alternative biome states has now extended to temperate and boreal regions (e.g. Scheffer et al. 2012b). Boreal ecosystems, long thought to be the archetypes of climate-control, are also experiencing re-evaluation of what really determines vegetation patterns. Here too the profound ability of organisms to alter their own environment, in this case soils, is challenging traditional views of climate controlling the major patterns of life on Earth. These are turbulent times for ecology.

CHAPTER 4

The nature of open ecosystems

4.1 Introduction

The nature of open ecosystems is surprisingly understudied, especially relative to forests occurring in the same landscape (Bond & Parr 2010). Among the reasons for this neglect are the long-held belief that they are secondary seral systems derived from forest by cutting and felling. The prejudice against 'anthropogenically disturbed' habitats can be so profound that biologists failed to recognize a major biodiversity hotspot in the grassy systems of the south-eastern USA until less than ten years ago (Noss 2012; Noss et al. 2015). Prejudices against open ecosystems are particularly strong where they also support regular fires. For much of the twentieth century, intensive efforts were made to suppress vegetation fires which were (and often still are) seen as human-caused environmental disasters, especially by the governing classes of many countries (Pyne 1990; Veldman 2016). Fire suppression, where effectively implemented, has resulted in the loss of pyrogenic open ecosystems and their biota and their replacement by closed forests (e.g. Noss 2012; Durigan & Ratter 2016).

In this chapter I introduce the plants and animals living in open ecosystems. To properly value these systems and their biodiversity we need answers to three key questions:

1. Are open ecosystems merely early successional phases derived from deforestation or are they instead stable 'old growth' systems of considerable antiquity?
2. What is the biodiversity value of open ecosystems and what would be lost if they were converted to other land uses including forestry plantations?
3. What are the traits of species living in open versus closed ecosystems and what do they reveal about the major processes that admit or exclude membership in the contrasting states?

4.2 Diversity, endemism, and the antiquity of open ecosystems

If an open ecosystem is merely an early successional stage following (anthropogenic) disturbance, then you would expect it to be short-lived and patchy in extent. Its biota, both plants and animals, would be expected to be poor in species, poor in endemics restricted to open habitat, and with low species turnover across physical habitat gradients. If an open patch was derived from forest, then the functional traits of species, especially woody plants, should be like those of forest habitats. If, on the contrary, open ecosystems are persistent and ancient, then you might expect selection for novel traits adapted to the open habitat and to the processes that maintain the open system. If open conditions have persisted over space and time, speciation may have occurred leading to species endemic to open habitats. If the open vegetation expanded into different climates and soils, with different disturbance regimes, then further speciation would be expected causing increased beta diversity (species turnover along habitat gradients) and greater total diversity of the open systems. Some examples will help to show how diversity and endemics have been used to infer the age and origins of open ecosystems.

Open Ecosystems: ecology and evolution beyond the forest edge. William J. Bond, Oxford University Press (2019). © William J. Bond 2019.
DOI: 10.1093/oso/9780198812456.001.0001

4.3 Tropical grassy biomes

4.3.1 The grasslands of Madagascar: anthropogenic or ancient?

The historical presence of humans, and human populations sufficient to open pristine forests, varies greatly around the world. In Africa, members of the genus *Homo* have been present for ~2.8 Ma and our species for ~200 000 years. Eurasia has the next longest record of hominin presence, followed by Australia (~45 ka), North America (~13 ka), and South America (~12 ka). Madagascar, the world's fourth largest island, was only settled in the last 2–4 ka, and New Zealand, only in the last 0.8 ka. Madagascar is a particularly interesting example. It is close to Africa, with which it shares many plant lineages, but also has many endemics. Its mammal and reptile fauna are strikingly different from Africa. Though famous for its forests, rich in endemic plant and animal species, more than two thirds of Madagascar is covered by C_4 grasslands. Numerous forest patches occur in the grasslands indicating that the climate can support closed forested ecosystems.

The usual explanation for such extensive grasslands is that catastrophic deforestation occurred after people arrived on the island and began burning the forest (Koechlin 1993; Kull 2000). In the last century, extensive deforestation has occurred in most regions because of fire and felling, especially in the high rainfall eastern side of the island (Green & Sussman 1990). But was massive earlier deforestation responsible for the extensive grasslands over the island? In 2017, the population was nearly 26 million, up from only 2 million early in the twentieth century. Through most of its history, large areas of Madagascar had very sparse human populations. A scientific debate began over the origin of the grasslands, with one side arguing that human populations were too sparse to account for their vast extent. Thus Grandidier (1898) stated that 'One has to admit that the [central plateau region of Madagascar] must have always been without trees, but not from the hand of man, since it was a vast deserted [landscape] of which not even 10% was populated'. The other side argued that the grasslands were artificial, as expressed by Perrier de la Bathie (1928): 'The formation of the prairies [of

Madagascar] is totally artificial and the result of fire. We would not insist on this except that some have said that the prairie, the malgache steppe, is a natural formation having existed since ancient times.' The anthropogenic argument 'won' and Madagascar gained the reputation for being one of the world's worst examples of deforestation with all its attendant consequences (Lowry et al. 1997; Kull 2000, 2004).

But there were problems with the argument. Burney (1987a,b) found grass pollen and charcoal in cores in lake sediments dating to thousands of years before human settlement. He noted that 'the apparent existence of open vegetation types, pyrogenic communities and shifting vegetation dominance in pre-settlement Madagascar is in direct contradiction to key aspects' of the deforestation hypothesis. Fisher and Robertson (2002) surveyed ant assemblages in forests and grasslands and found similar numbers of species in both habitats. Most (18) of the grassland ant species were found only in grasslands and only five species occurred in both habitats. They concluded that open habitat must be far older than suggested by proponents of a deforestation origin. Bond et al. (2008) reviewed the evidence for an endemic grassland biota in Madagascar. Besides the grasses themselves, they noted the presence of grassland forb species endemic to Madagascar, and numerous *Erica* species, twice as diverse as all of Europe (>40 species versus ~20). Ericas grow in open habitats with regular fires and cannot survive in forest shade. Along with plant species, there are birds, reptiles, ants, and giant earthworms endemic to grassland and shrubland habitats. An endemic genus of grass-feeding termites, *Coarctotermes*, is restricted to the Madagascan grasslands. Bond et al. (2008) concluded that the rich, endemic grassland biota indicated Madagascar's grasslands were ancient and part of the global expansion of C_4 grassy biomes from the late Miocene (Chapter 5).

Recent studies of the grasses that make up the Madagascan grasslands support ancient origins. In the first studies of grasses, other than bamboos, since the 1960s, Vorontsova et al. (2016) reported a total of 541 grass species of which 40 per cent are endemic making it the second richest island for grass endemics in the world (New Zealand is first). Grassland endemics are distributed across the island suggesting that grasslands were formerly widespread long before human settlement. Vorontsova et al. (2016) pointed out that the proportions of species in the most important subfamilies and tribes in Madagascar are like those of East Africa, a region of extensive savannas on the mainland. The only exception is the bamboo lineage, which has far more species in Madagascan forests than in Africa. Vorontsova et al. (2016) concluded that the open-habitat grass flora is ancient, on a par with Africa, and certainly far older than human settlement.

While there is little doubt that Madagascan grasslands are ancient, there is still uncertainty as to its pre-settlement areal extent. For example, there is a puzzling lack of isotopic evidence for grazing by the endemic vertebrate fauna (Godfrey & Crowley 2016). Is this because grasslands, though present, were too small to select for grazers? Or did herbivores prefer more nutritious C_3 plants if they could access them? Is the apparent C_4 avoidance an artefact of small sample sizes of fossil material? In some ways, Madagascan nature is in a time warp, giving us a glimpse of the ecological repercussions of the C_4 grass revolution when savannas first began to sweep across the tropics from ~7 Ma ago. The history of the spread of grasslands deserves much more study on the island, hopefully less charged with rhetoric than the controversies of the early twentieth century. Because settlement is so recent (<5000 years), it could be the first large land mass in which we can quantify prehistorical human impacts on the extent of tropical forest.

4.3.2 Southern grasslands of the USA

A second example of delayed recognition of the antiquity and biodiversity of grasslands is from North America. The Atlantic and Gulf Coastal Plain Floristic Province of the USA includes grasslands, savannas, and broad-leaved forests. Longleaf pine (*Pinus palustris*) is common and widespread in the savannas, producing a sparse canopy overlying a frequently burnt C_4 grassy understorey. As for Madagascar, and many other tropical grassy biomes, the southern grasslands and pine savannas were interpreted as products of deforestation, here attributed to burning by Native Americans (e.g. Mann

2005). Active fire suppression after European settlement led to replacement of the pine savannas by closed broad-leaved forests. Today a mere 2–3 per cent of the pre-European area of the longleaf pine ecosystem still exists (Noss 2012).

Recently Noss and colleagues (Noss 2012; Noss et al. 2015) have pointed to the remarkable plant species richness of the southern grasslands, including the pine savannas, of the USA. The floristic province contains some 6170 plant taxa, nearly a third of the flora of North America, of which 1748 taxa (28.3 per cent) and 51 genera are endemic to the region. Nearly 1000 of the endemic plant species occur in the grasslands and pine savannas. Like the savannas of the southern hemisphere, the coastal plain grasslands escaped glaciation, retaining their biota through the Pleistocene. In contrast, the Prairie Region of the Great Plains was only assembled after the ice retreated, and has less than a tenth of the endemic species (~ 87) and no endemic genera. The southern grasslands, far from being a recent product of anthropogenic activity, turn out to be ancient, remarkably rich in species, and now belatedly recognized as a global biodiversity hotspot (Noss et al. 2015). The embarrassingly late discovery of the hotspot is, according to Noss et al. (2015) the result of cultural biases that have blinded scientists to the existence of this ancient species-rich system. The anthropogenic origin narrative was so persuasive that people believed in it, forgetting it was a hypothesis and, like any others, needed testing. There are obvious parallels with other non-forested regions of the world, especially those where fires are common.

4.4 North temperate floras

Compared to the tropics, the antiquity of open ecosystems in north temperate regions seems even less well explored. Yet species-rich open habitats in 'forest' climates also occur in temperate regions. The usual story, again, is that they were created by deforestation of a vast primeval forest. Though challenged, most notably by Vera (2000, 2002), this view still seems to prevail. In Europe, untransformed grassland habitats are rare and often conserved for their unique array of plant and animal species. Usually described as 'semi-natural', some authors interpret them as relics of the last glacial period. For example, Weigl and Knowles (2014) pointed to similarities in the attributes of 'balds' in the forests of the eastern USA and *poloninas*, vast grasslands of the Eastern Carpathian Mountains in Poland, Ukraine, and Slovakia. The balds are grassy, upland habitats surrounded by a 'sea' of forest. They do not burn but did support (and some still do) populations of deer and caprics (sheep and goats) which browse saplings of forest trees colonizing the balds. Balds have a flora distinct from the neighbouring forests, pointing to some antiquity of the habitat. Weigl and Knowles (2014) suggest that these open ecosystems are legacies of the Ice Age, maintained under current climates by browsing.

A similar argument has been put forward for the extensive heathlands and moorlands of the Scottish Highlands (Fenton 2008) though others have argued against it (Bennett 2009; Peterken 2009). While parts are too cold for forest growth, large areas of the Highlands can support closed conifer forests, as shown by forestry plantations, especially of Sitka spruce (*Picea sitchensis*) and patches of broad-leaved forests (Plate 2). The native Scots pine, *Pinus sylvestris*, has an open canopy and supports a shade-intolerant understorey. Fenton argued that these open ecosystems are also relics of the last glacial maintained by browsing. That they can revert to forests when browsers are excluded is apparent on islands in the lochs, and behind deer and sheep-proof fences (e.g. Palmer & Truscott 2003).

A key question for the origin of north temperate open ecosystems is: where does the open habitat flora and fauna come from? Are the species drawn from source areas too cold or too dry for forests? Or are they native to open habitats within the general forested matrix? If the latter, then the open habitats are of much greater antiquity than the opening of the forests by humans in the Holocene.

4.5 A global perspective: open ecosystem biodiversity hotspots

Contributions of open ecosystems to world biodiversity can be seen in their representation in global biodiversity hotspots. Hotspots were identi-

fied by Myers et al. (2000) as a tool for setting global conservation priorities by pointing to areas particularly rich in species. Hotspots are biogeographic regions with >1500 endemic plant species and with significant loss of primary habitat (<30 per cent remaining). Thirty-five hotspots have been identified (Mittermeier et al. 2011) accounting for 17.3 per cent of the Earth's land surface but more than 77 per cent of plant species and 43 per cent of vertebrates (Marchese 2015). Though the concept has been criticized (see Marchese 2015), it has contributed significantly to world conservation efforts.

Among the 'hottest' of hotspots are several regions dominated by pyrophilic open vegetation (Table 4.1). They include the savannas of Brazil (cerrado), rich in plant species, including many endemics, and also rich in birds, reptiles, and mammals. Mediterranean-climate regions include the major biodiversity hotspots of the Mediterranean basin itself, the California floristic province, the Cape floristic region of South Africa, and the ecologically similar region of south-western Australia, all dominated by flammable shrublands (Table 4.1). Table 4.1 also lists some additional regions of high biodiversity flammable grassy vegetation: the North American coastal plain recently listed as an additional global hotspot (Noss et al. 2015); the campo rupestre of Brazil, an extraordinarily species-rich flammable grassland in the high rocky mountains of Brazil (Silveira et al. 2016; Fernandes 2016); the Campos Sul (the southern grassland biome of Brazil forming mosaics with closed forest) (Overbeck et al. 2007); and the physiognomically similar grassland biome of South Africa, also an upland grassland rich in species (Gibbs Russell 1987; Bond & Parr 2010).

Table 4.1 lists some closed-forest biodiversity hotspots for comparison, including the remarkable Atlantic forests of Brazil with twice the number of plant species and plant endemics in almost half the area of cerrado. Other hotspots, especially mountainous ones in Eurasia (e.g. Caucasus, Irano-Anatolian, Mountains of Central Asia), include significant areas of species-rich open ecosystems, in mosaics with forest. Mammals, rather than fire, appear to be the major consumers maintaining open vegetation in these northern hemisphere upland hotspots.

Table 4.1 Biodiversity hotspots ranked by plant species number. Open flammable hotspots include Mediterranean climate shrublands and cerrado, a savanna. Tropical and temperate forests are listed for comparison. Additional grasslands in North America, Brazil, and South Africa are biodiverse but not yet listed as global hotspots. O = occupant, E =endemic. Hotspot data from Mittermeier et al. (2011). Grassland data from, respectively, Noss et al. (2015); Silveira et al. (2016); Gibbs Russell (1987); Overbeck et al. (2007). Regions are ranked by total plant species.

Hotspot	Area	Plants		Birds		Reptiles		Amphibians		Mammals	
Open flammable hotspots	1000 km²	O	E	O	E	O	E	O	E	O	E
Med Basin	2085	22 500	11 700	497	32	228	77	91	41	216	27
Cerrado	2036	10 000	4400	605	16	225	33	205	34	300	10
Cape Fl Region	90	9000	6210	324	6	100	22	47	16	109	0
SW Australia	303	5571	2948	285	10	177	27	32	22	55	13
Calif Fl Province	294	3488	2124	341	8	69	4	54	27	141	15
Forest hotspots											
Atlantic Forest, Brazil	1250	20 000	8000	936	148	306	94	516	323	312	48
Forests, East Australia	222	8257	2144	632	28	321	70	120	38	133	6
Valdivian-Chile forest	248	3892	1957	226	12	41	27	44	32	69	19
Additional Grasslands											
N.Am. Coastal Pl all taxa	1130	6200	1816	NA	NA	291	113	122	57	306	114
full spp				*274*	*6*	*177*	*50*	*105*	*45*	*148*	*9*
Campo rupestre, Brazil	66	5011									
Sth Afr grassland biome	112	3800									
Campos Sul, Brazil	137	3000-4000									

4.5.1 Forests versus savannas: an intercontinental comparison

Until recently, there were no large-scale comparisons of the diversity in closed forest versus open ecosystems. A recent study has compared savannas versus forests (Murphy et al. 2016). As we have seen, savannas have generally been viewed as less diverse than forests and assumed to have less turnover along habitat and geographic gradients. However, at the continental scale, savannas have respectable levels of diversity, particularly in animal groups. Global compilations of species richness data, organized into ecoregions (Olson et al. 2001; Bailey 2014) allow comparison of largely forested ecoregions with largely OE ecoregions (Figure 4.1; Murphy et al. 2016). Forests are richer than savannas in all four biogeographical regions. However, when adjusted for rainfall and latitude (diversity increases with rainfall and towards the Equator), savannas have equivalent richness to forests for mammals, birds, and amphibians. Plant richness,

however, is nearly half that of forests in the savanna floras. Savannas also consistently had fewer range-restricted species than forests for the faunal groups (Murphy et al. 2016). As noted by Bond and Parr (2010), the comparative neglect of studies of the diversity of tropical grassy biomes seems to have far more to do with cultural blind spots of researchers than with any real differences in the diversity of forests versus OEs.

4.6 The diversity of open ecosystems

With growing recognition of the antiquity of open vegetation, and a definition of OEs that emphasizes shade-intolerance of their major growth forms, rather than the density of trees, new centres of diversity are being 'discovered'. With discovery comes renewed interest in diversity of the biota, as in the case of Madagascar. Systems once considered anthropogenic and therefore uninteresting are proving to be fascinating. In Brazil, for example, the herpetofauna of the cerrado biome was considered monotonous

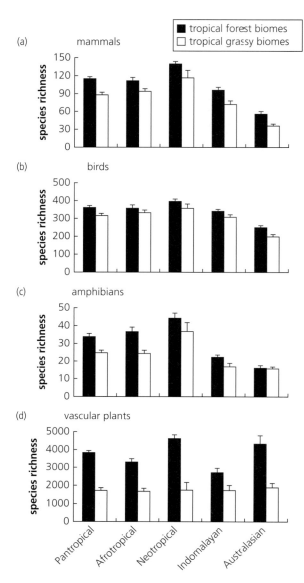

Figure 4.1 Comparison of species richness between tropical forest and tropical grassy biomes in each biogeographic realm, fcr key vertebrate groups: (a) mammals, (b) birds, and (c) amphibians, as well as (d) vascular plants. The means are calculated from the values for each tropical ecoregion. The error bars indicate standard error of the mean (of ecoregions). From Murphy et al. (2016).

with negligible turnover across its 2 million km² area, as you might expect if it was recently cobbled together in the wake of human deforestation. But new studies are revealing remarkable diversity and high speciation of lizards (Noguiera et al. 2011;

Werneck 2011) consistent with new studies indicating late Miocene origins of cerrado (Simon et al. 2009; Chapter 5). In Madagascar, genetic evidence of a population bottleneck of a rare lemur has been interpreted as a legacy of forest fragmentation as grasslands spread long before humans settled on the island (Quéméré et al. 2012). In India, a study of the lizard genus, *Ophisops*, endemic to open-habitat grasslands, was conducted to help date its origins (Agarwal & Ramakrishnan 2017). As for cerrado, the species diversity of the genus turned out to be grossly underestimated while a dated molecular phylogeny revealed an 8-fold increase in diversification rate of one clade during the global C_4 grassland expansion (Chapter 5). Among the new areas of discovery is recognition of the savannas of Asia (Ratnam et al. 2016), long misnamed as 'dry forests' and mismanaged by supressing fires. Asking new questions on the diversity, endemism, and origins of the biota of open vegetation is a very necessary awakening of scientific curiosity about these systems.

4.7 Functional traits in a trophically structured world: a framework

If open ecosystems are maintained by consumers, you would expect their plant traits to diverge from those of closed forests. There is a very large literature on plant traits, especially those relating to resource use (Díaz et al. 2016). The leaf economic spectrum (Wright et al. 2004) and the gradient from conservative to acquisitive resource use (Díaz et al. 2004) have been widely explored. However, traits related to resource use may not discriminate between species of open and closed vegetation (Charles-Dominique et al. 2015b). Instead, quite different traits are emerging as indicators of these states and of the consumers that maintain open states. These studies are new and novel traits are still being discovered.

Chapter 3 introduced the concept of a green, black, and brown world. The green world is that of closed-canopy ecosystems where trees are at equilibrium with climate. Black and brown worlds are open ecosystems, with lower biomass than their climate potential, and with a shade-intolerant understorey. In black worlds fires occur regularly and shape vegetation structure and composition. In brown worlds, mammals are the major consumers maintaining open

Figure 4.2 Tree architecture is very different in frequently burnt versus heavily browsed savannas. Left: *Acacia karroo* (=*Vachellia karroo*), with pole-forming saplings, Right: *Acacia grandicornuta*, with cage-like saplings in heavily browsed savanna.

structure and plant composition. If the three worlds have ancient origins, then plants in each should diverge from the other in functional traits. Plants living in open versus closed worlds would be expected to differ in shade tolerance, drought and heat responses, and, possibly, nutritional strategies (see Chapter 1). In open ecosystems, black world traits would be expected to diverge from brown world traits (Staver et al. 2012). Traits suitable for each world are likely to diverge because of trade-offs in resource allocation for different circumstances. For example, dense branching may help reduce losses to browsers by forming a cage protecting foliage and stems. However, dense branching is hopeless for rapid height growth in a forest gap or for escaping flames in flammable ecosystems (Figure 4.2). If different trait syndromes characterized each of the three worlds, the traits could be used to identify the dominant processes structuring an assemblage. Plants invading a system with different traits could signal regime shifts among the three states (Bond et al. 2001).

4.7.1 Green versus black and brown world traits (closed versus open ecosystems)

Grime (1979) introduced a trade-off triangle to characterize species along gradients of competition, stress, and disturbance. In consumer-dominated

systems, it is possible to ordinate species in trait space according to whether they are best fitted to green, brown, or black world (Grime 2006). Figure 4.3 (Plate 7) shows the basis for such a schema developed by Tristan Charles-Dominique for a southern African savanna/forest mosaic system (Charles-Dominique et al. 2015b, 2018). Its generality is being tested in other parts of the world and additional traits are being explored. An interesting finding, thus far, is that traits used for locating species along productivity gradients, such as SLA, do not clearly distinguish closed and open ecosystems in samples explored to date.

4.7.2 Black world traits

Where fire is the consumer, there is a common suite of traits discriminating between open (flammable) and closed (forest) vegetation. The traits differ depending on the fire regime, defined as the patterns of frequency, season, type, severity, and extent of fires in a landscape (Chapter 7; Gill 1975; Bond & Keeley 2005). There are two major types of fire that select for different suites of traits: crown fires and surface fires with varying combinations and permutations (Figure 4.4). Crown fires consume all above-ground biomass, are characteristic of woody fuels, burn at decadal+ intervals, with high fire intensity, and burn all foliage,

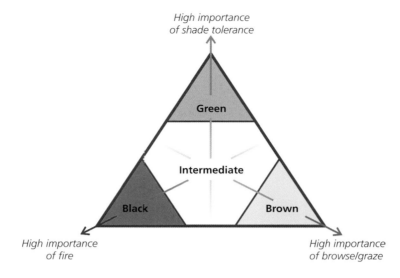

Function	Trait	Green *Surviving shade*	Black *Surviving fire*	Brown *Surviving herbivory*
Growth	Height growth	rapid	intermediate	slow
	Branch mass/Leaf area	Low	High	High
	LAI	High	Low	Low
	SLA	Variable	Variable	Variable
Fire	Bark growth rate/thickness	Low	High	Variable
tolerance	Bud insulation	Low	High	Variable
	Accessory buds	Very low	High	Medium
	Hydraulic fire resistance	Low	High	Low
Herbivore	Structural defences			
tolerance	Spinescence	Low	Variable	High
	Ramification	Low	Low	High
	Bite size	High	Medium	Low

Figure 4.3 Green, black, and brown world for savanna versus forest woody plants represented as a trade-off triangle. The table lists traits grouped under those supporting growth, fire, and herbivore resistance. Fire traits are for a surface, grass-fuelled fire regime. Herbivore resistance traits are for ungulate herbivores. See Charles-Dominique et al. (2015b, 2018) for worked examples and methods used to classify trait space for tree species. (See Plate 7).

whether green or dead. Crown fires are characteristic of shrublands, including those of Mediterranean-type climates, heathlands, and some conifer woodlands. Surface fire regimes are fuelled by grasses, low shrubs, and litter. They include the vast grass-fuelled savannas of the tropics but also litter-fuelled fires in some conifer and other woodlands, including boreal forests with lichen understories. Tropical savannas

have a highly productive grassy layer accumulating flammable fuels that burn several times in a decade or, in some areas such as the western Serengeti, twice in a year. Savanna fires are fast moving, of low intensity, with most damage to stems of plants in the flame zone (1–3m). Surface fires can flare up where fuels form a ladder, causing locally intense burn-outs. Mixed fire regimes of both surface and

Figure 4.4 Surface fires and crown fires select for very different traits. Left, surface fire in eucalypt savannas sustained by grass fuels. Taller plants escape direct fire damage. Right: crown fire in Californian chaparral. The shrub canopies burned off and they are resprouting from the base or from seeds stored in the soil.

crown fires are the norm for some conifer woodlands, such as those in the western USA, and eucalypt-dominated woodlands in Australia.

4.7.3 Grass-fuelled surface fire regimes

Bark. There is a large literature on fire regimes and associated traits (e.g. Scott et al. 2014 for an introduction) and the reader is referred to this literature for more information. Here I am concerned with contrasting traits in open and closed vegetation. Since the key traits are quite different for different fire regimes, it is necessary to start by defining the fire regime of the vegetation of interest. For frequently burnt savannas, there is a marked floristic turnover of woody plants from open to closed vegetation closely linked to functional turnover. Bark thickness is one of the key differences in green versus black world trees (Cole 1986; Pausas 2015; Lawes et al. 2013; Charles-Dominique et al. 2017a). Rapid bark growth rate in juvenile plants discriminates forest from savanna tree species with the most rapid growth rates where fires are most frequent (Figure 4.5). Clonal sprouting species ('root-suckering' plants) are exceptions to the general rule because their perennating organs are insulated below ground. Among forest trees, thick-barked trees do sometimes occur but

the selective benefits are not well understood (Rosell 2016).

Buds. Besides thick bark, bud insulation is also very different in the woody plants of black versus green world, influencing patterns of resprouting after injury (Clarke et al. 2013). Burrows (2002; Burrows et al. 2010) initiated studies on the importance of bud insulation for fire survival by reporting that eucalypts owe their remarkable ability to tolerate fire damage because their meristematic tissue is deeply buried under the bark. Because of bud protection, eucalypts can regenerate rapidly from fire-damaged canopies by epicormic sprouting (Figure 4.6). Subsequent studies have shown that eucalypts are not unique in this respect and many trees, especially in savannas, sprout epicormically from deeply embedded buds (Charles-Dominique et al. 2015a). The degree of protection of buds can be categorized (Figure 4.7; Plate 7) and these categories of protection are markedly divergent in forests versus savanna trees with the latter far better protected. The consequence is that, after the canopy is scorched, savanna trees recover their canopy rapidly from epicormic sprouting whereas forest trees are killed or sprout from the base (Charles-Dominique et al. 2015a; Pausas & Keeley 2017).

Hydraulics. There may be an additional twist to the story. For many decades, it has been assumed

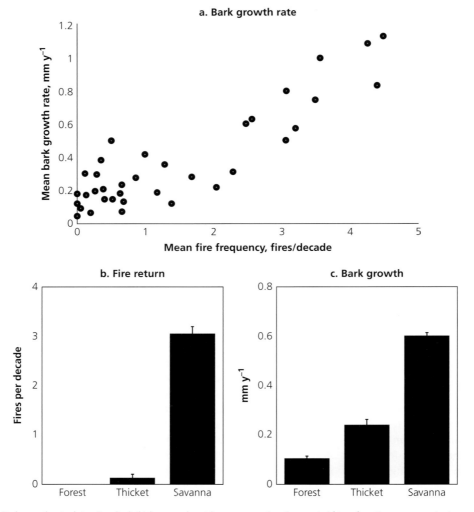

Figure 4.5 Bark growth rate determines bark thickness and post-burn recovery time in a mesic African forest/savanna mosaic. A. growth rate increases with increasing fire frequency. Each point is a species. B. Fire return in closed (forest, thicket) versus open ecosystems. C. Mean bark growth rates of trees in the three biomes. From Charles-Dominique et al. (2017).

that trees die after fires because the cambium is killed. Jeremy Midgley and colleagues (2011) argued that this mechanism is far too slow to explain rapid post-burn stem death. They proposed that fire caused embolisms in the conducting tissue and argued for a 'hydraulic death' hypothesis instead (Midgley et al. 2011). Subsequent studies have supported the hydraulic death argument with increasingly detailed analyses of heat-induced cavitation by plastic deformation of the xylem vessels or high evapora-

tive losses as the fire front approaches (Michaletz et al. 2012; West et al. 2016; Michaletz 2018). These new ecophysiological studies suggest novel hydraulic traits by which open-habitat trees may differ from forest trees. Greater resistance to heat-induced cavitation may, along with differences in bark and bud traits, allow faster canopy recovery from canopy scorch in fire-resistant trees.

Juvenile traits. Green and black world trees diverge in sapling traits. In savannas, saplings contend with

Figure 4.6 Trees with well-insulated buds and resistance to heat-induced cavitation can recover their canopy rapidly after a fire. In this example, the Eucalyptus tree is sprouting epicormically, contrasting with pine trees killed by the fire.

Dormant bud protection

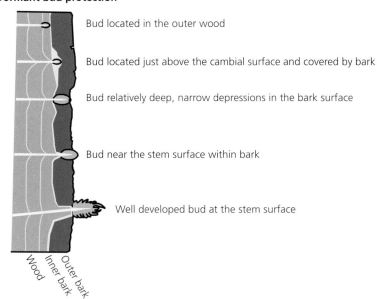

Bud located in the outer wood

Bud located just above the cambial surface and covered by bark

Bud relatively deep, narrow depressions in the bark surface

Bud near the stem surface within bark

Well developed bud at the stem surface

Wood
Inner bark
Outer bark

Figure 4.7 Buds vary in the location in relation to bark. The relative insulation from fire injury varies from greatest at the top of the figure to least at the bottom. Figure from Charles-Dominique et al. (2015a), with permission. (See Plate 7.)

frequent fires which kill stems in the flame zone ('top-kill'). Savanna trees commonly have saplings with swollen underground storage organs, or lignotubers, often packed with starch, which facilitate rapid resprouting after fires (Wigley et al. 2009; Simon et al. 2009; Schutz et al. 2009; Hoffmann et al. 2003, 2004; Simon & Pennington 2012) (Figure 4.8). Stems tend to be sparsely branched and vertically oriented, especially in trees that only mature above the flame zone (Archibald & Bond 2003). The saplings ('gullivers'; Bond & van Wilgen 1996) of such trees can spend decades in the flame zone, topkilled by fires, resprouting, then topkilled again. Only those few saplings that can grow above the flame zone between fires will grow into mature trees which are relatively fire-proof (Higgins et al. 2000). Underground storage organs are rare in forest trees, especially relative to savanna species (Hoffman et al. 2003).

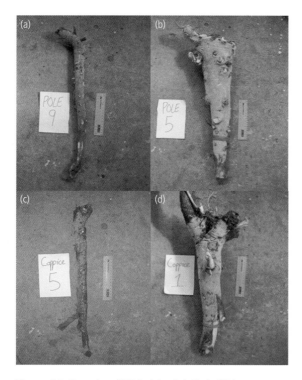

Figure 4.8 Lignotubers (USOs) of *Acacia* (=*Vachellia*) *karroo* sampled in a frequently burnt mesic savanna. Figures a,b are smallest and largest USOs from unburnt plants; c and d are smallest and largest from post-burn resprouting plants. Lignotuber dry weight (unburnt) had a mean of 538 ± 110g with TNC concentration of 32.8 ± 4%, and a TNC pool of 150 ± 19 g. From Wigley et al. (2009).

4.7.4 Underground trees: geoxylic suffrutices

A very interesting question in savannas is 'why be a tree?' With frequent fires, saplings may be trapped in the flame zone for decades. Why not allocate resources to early reproduction in the juvenile phase, even if the cost is never growing out of the flame zone? Many species have done just that by taking on a shrub growth form which flowers and fruits within the flame zone (though some may grow above it with long intervals between fires). Even more extreme selection for height of reproduction is seen where fires are frequent and growth rates are slow because of limited soil nutrients, seasonal waterlogging, or periodic frost (Simon & Pennington 2012; Maurin et al. 2014). These habitats support 'underground trees' (= 'geoxylic suffrutices'). These are woody plants whose branches are subterranean with the foliage being borne on short-lived above-ground shoots (Figure 4.9; Plate 8). Underground trees, along with many savanna shrubs, have sister species that are savanna or forest trees. They are common in humid savannas in Brazil (cerrado) and Africa (especially 'miombo') and do not occur in forests. Underground trees can have large canopies up to 50 m diameter and reach great ages—with plants of *Jacaranda decurrens* species in Brazil estimated to be 3800 years old (Alves et al. 2013). Their demography is unknown and therefore the trade-offs in adopting this life history are also unknown. They do seem to offer the benefit of large tree size in having a large canopy capable of bearing many flowers and fruits, contrasting with shrubs which have far smaller canopy area. However, they may take even longer than trees to build their large canopy size, thereby limiting their success.

As noted earlier, much of the species richness of grassy systems is among the herbaceous plants. Forbs with large underground storage organs are prominent in 'old growth' tropical grassy systems (Veldman et al. 2015a). The evolutionary origin of the forbs has not been studied at the time of writing and, other than finding that the forbs are eliminated by shading, no formal trait comparisons with comparable growth forms in forests have been made.

4.8 Crown fire systems

Unlike savannas, there are few comparisons of plant traits in green versus black world for crown

Fire   Resprouting   Flowering   Persist as short stems
                                                                    between fires

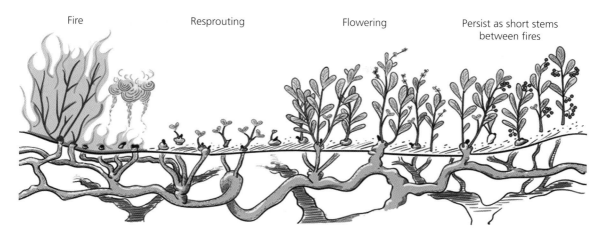

Figure 4.9 'Underground trees' (geoxylic suffrutices). Some woody species with tall tree relatives form "underground trees", with massive below-ground branches supporting short above-ground stems that resprout rapidly after a fire (Top: from Bond (2016), *Science*). The example below (within the white line) has a much larger canopy spread than the co-occurring small trees and shrubs; in cerrado, Brazil. (See Plate 8.)

fire systems. However, plant traits in crown fire regimes of black world have been intensively studied, especially in Mediterranean-type shrublands (for reviews see Bond & van Wilgen 1996; Pausas et al. 2004; Keeley et al. 2012). There are many instances of convergence in fire response traits in Mediterranean-type shrublands despite divergent

origins of the flora in the different regions. Unlike surface fire regimes, few woody plants have thick bark since above-ground stems are burnt in most fires. Many woody plants resprout from the base, some from above-ground stems where buds survive the fire, while others are routinely killed (nonsprouters or 'seeders'). For a thorough account see

Keeley et al. (2012); Clarke et al. (2013). A similar array of responses is observed when fires burn into closed forests. Tree species may be non-sprouting, basal sprouting, or, rarely, epicormic sprouting (Bond & Midgley 2001; Pausas & Keeley 2017). However, comparative studies of open and closed vegetation in the same region are few.

Reproductive traits. While vegetative responses to fire may be similar, reproductive responses are very different in green versus black world crown fire systems. Post-burn recruitment is common in black world plants cued by fire-stimulated flowering, seed release from insulated structures (serotiny), and fire-stimulated (heat shock, smoke) seed germination. A useful indicator of a crown fire system resilient to fire is the presence of seedlings in the first year or two post-burn. In closed forests, where fires are the exception, fire-stimulated recruitment is also the exception.

It is perhaps surprising that fire-stimulated recruitment is rare or absent in woody savanna species. Fires may be too frequent and grass recovery too rapid for sufficient benefits to delay germination to post-burn timing. However, well-studied examples are few so generalizations are premature. Studies in Brazilian cerrado, especially the grassier variants growing on nutrient-poor quartzite-derived soils, show marked flowering responses to fire in both grasses and forbs (Fidelis & Blanco 2014). Among fire-stimulated flowering species are the sedge *Bulbostylis*, which flowers within a few hours of being burnt, and the astonishing Eriocaulaceae species, the 'rocket' plants, with inflorescences like a frozen bursting firework (Figure 4.10; Plate 8). A parallel response of fire-stimulated flowering has been reported for mesic African grasslands (Zaloumis & Bond 2016) but the phenomenon requires more study.

4.9 Old-growth grasslands

How do you tell whether a patch of vegetation is ancient or secondarily derived from deforestation? At a superficial level, a tropical grassland derived from

Figure 4.10 Many members of the Eriocaulaceae, such as this population of *Paepalanthus urbanianus* Ruhland ('rocket plants'), flower en-masse after fire in the cerrado of Brazil. Fire-stimulated flowering is common in many herbaceous species in crown and surface-fire regimes in different parts of the world. (See Plate 8.)

Table 4.2 Comparison of species richness for different growth forms in natural versus derived (secondary) savannas. SE is standard error (values in italics are range). Data are species number per plot size. Plot size was constant within the Brazil study (Veldman & Putz 2011) and within the three South African study sites (Zaloumis & Bond, 2016). The column n/d is the quotient of species number in natural:derived plots for that growth form.

	natural (n)	SE	derived (d)	SE	n/d
BRAZIL					
grasses	5.7	0.6	1.7	0.3	3.4
savanna trees	2.5	*1.2–5.8*	0.4	*0–7.2*	6.3
forest trees	0.6	0.1	1.3	0.3	0.5
% savanna trees	83		22		3.8
SOUTH AFRICA					
MNR (Montane grassland)					
forbs	18.5	1.07	6.4	0.29	2.9
woody	2.4	0.2	0.2	0.07	12.0
geoxylic	1.7	0.23	0.1	0.05	17.0
graminoid	6.5	0.43	5.1	0.39	1.3
BKNR (Montane grassland)					
forbs	30.2	2.08	5.4	0.66	5.6
woody	0.3	0.23	1.1	0.18	0.3
geoxylic	0.4	0.08	0.1	0.05	4.0
graminoid	12.6	0.92	2.3	0.13	5.5
ESI (Coastal Belt grassland)					
forbs	18.5	1.07	6	0.45	3.1
woody	2.4	0.2	2.7	0.27	0.9
geoxylic	1.7	0.23	0.1	0	17.0
graminoid	6.5	0.43	5.4	0.35	1.2

deforestation may seem very similar to a grassland that has remained a grassland for thousands of years. In a study of a forest/savanna mosaic in lowland Bolivia, Veldman (2016; Veldman and Putz 2011) compared attributes of primary savanna to savanna derived from deforestation of deciduous tropical forest. 'Primary' savanna had remained savanna throughout the historical record. The primary and derived savannas were structurally similar with very similar grass and tree cover. However, they diverged markedly in grass and tree species composition and in species richness (Table 4.2). Natural savannas had entirely different dominant grass species from derived savannas where naturalized African grasses were prominent. Trees that survived in the derived savanna were mostly remnant individuals from the forest (78 per cent of individuals) whereas trees in the natural savanna were overwhelmingly of savanna origin (only 17 per cent of forest affinity).

Similar large differences in grass species composition were reported by Zaloumis and Bond (2016) for lowland and montane South African grasslands (Table 4.2). The grasslands had been afforested with conifers, the trees felled, and, 20 to 40 years later, the composition of the secondary grasslands was compared with adjacent grasslands with no history of afforestation. The secondary grasslands had completely different grass species composition and were clearly visible in the dry season from their white colour (*Eragrostis* spp.) contrasting with natural grasslands which cure to a reddish hue (dominated by Andropogoneae). The grasslands had entirely different forb floras with three to five times more species in natural versus derived grasslands. These forbs were dominated by long-lived perennials with large underground storage organs (USOs) which resprout and grow rapidly after fire before being suppressed by grasses (Figure 4.11; Plate 9). Primary grasslands also contained 'underground trees' (geoxylic suffrutices), which had mostly been eliminated in secondary grasslands. As in Bolivia, primary and secondary grasslands appear structurally similar. However, closer observation showed a reduction in below-ground forb biomass from 3.2 kg m^{-2} to 0.1 kg m^{-2} at one site and 1.3 to 0.01 kg m^{-2} at a second site (Zaloumis & Bond 2016).

Studies such as these indicate the need to distinguish ancient from secondary grassland. Veldman et al. 2015a, have defined the features of 'old-growth' grasslands and noted their analogies to 'old growth' forests (Table 4.3). They possess special attributes quite different from secondary grasslands recovering from ploughing, afforestation, and other major disturbances. Unlike forests, old-growth grasslands cannot easily be identified from space by simple structural features. But such markers must be found to help identify the areal extent of ancient grassy systems by remote sensing to reduce wanton transformation.

4.10 Brown world traits: herbivory responses

4.10.1 Woody plants

Insect herbivory and plant defences have been widely studied but vertebrate herbivory and plant defences are relatively poorly known in comparison. One of the best indicators of concentrated vertebrate herbivory is the presence of structurally defended plants. Whereas chemical defences mostly work after ingestion by the animal, structural defences influence feeding behaviour and reduce intake before ingestion. Spines are by far the most common woody plant defence in most, but not all, regions (Figure 4.12). Spines are an interesting defence: spiny plants often have highly palatable foliage, are a major dietary item for the animals that feed on them, yet frequently dominate

Table 4.3 Characteristics of many old-growth grasslands, savannas, and open woodlands. (Modified from Veldman et al. 2015a.)

Ecosystem-level characteristics	Life history, functional characteristics of old-growth indicators
Species assemblages not present in young 2° grasslands	Slow growing
High herbaceous layer species diversity	Long-lived
High point (1m^2) scale species richness	Strong resprouters
Presence of endemic species	Poor recruitment from seed
Transient seedbanks	Poor colonizing ability
Persistent bud banks	Large investment in underground storage organs
High ratio of herbaceous to woody species	Clonal growth
High below-ground biomass	High root: shoot ratio
Little accumulation of litter or duff	Fire-stimulated flowering and fruiting
Open, discontinuous tree canopies	Fire-tolerant, thick-barked trees
Factors that maintain biodiversity	**Causes of degradation**
Frequent surface fires	Fire suppression/exclusion
Vertebrate herbivory, including livestock	Woody encroachment
Soil disturbance by digging animals	Invasive species
Shallow soils; nutrient-poor soils	Atmospheric N deposition
Seasonal drought, seasonal flooding	Anthropogenic soil disturbance
	Heavy grazing
	Elevated atmospheric CO$_2$
Land uses incompatible with old growth	**Land uses compatible with old growth**
Surface mining or quarrying	Low intensity grazing
Tillage agriculture	Timber, fuelwood extraction
Plantation forestry	
Intensive pasture management	
Prolonged high intensity grazing	

vegetation where browsing pressure is high. Spines function by reducing bite size of the browsing animal, thereby reducing food intake rate and driving the animal to move away to seek more rewarding targets (Cooper & Owen-Smith 1986). From the plant perspective, spines vary in function from

Figure 4.11 Tropical grassy biomes are rich in forb species many of which have large underground storage organs and dwarf woody species with clonal spread (c). This collection is from cerrado in Brazil (a–c), and grasslands in South Africa (d–f). a. *Psidium grandifolium*, Brazil, b: *Pfaffia jubata*, c: *Bauhinia dumosa*, d: USOs of *Gnidia*, above, *Clutia*, below; e: *Raphionacme*; f: *Eriosema* emerging from a sod of grassland. Courtesy of a,b, Giselda Durigan, c, Alessandra Fidelis, d–f, Nick Zaloumis. (See Plate 9.)

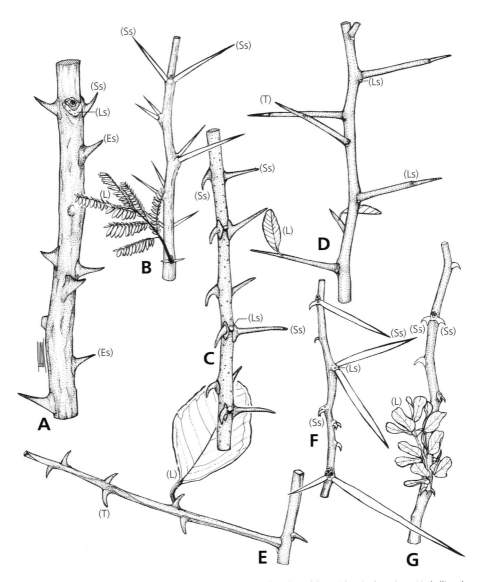

Figure 4.12 An assortment of spines from African trees. (A) Prickles: *Zanthoxylum davyi*. (B) Straight stipular spines: *Vachellia robusta*. (C) Straight stipular spines and stipular hooks: *Ziziphus mucronata*. (D) Straight thorns: *Gymnosporia harveyana*. (E) Hook thorns: *Scutia myrtina*. (F) Straight stipular spines and stipular hooks: *Vachellia tortilis*. (G) Stipular hooks: *Senegalia nigrescens*. Es, epidermic spine; L, leaf; Ls, leaf scar; Ss, stipular spine; T, thorn (i.e., branch with a sharp tip). From Charles-Dominique et al. (2016).

those that reduce leaf intake to those that reduce branch removal (e.g. Gowda 1996; Charles-Dominique et al. 2017b). The latter is very important in controlling height growth and therefore the probability of juvenile recruitment into larger size classes. Spiny plants are rare in shaded forest understoreys for reasons that are not well understood. The cost of carbon allocation to building

spines might be too great in the shade when traded off against reduced allocation to height growth. Another possibility is that architectural properties such as short shoots and short internodes may be incompatible with life in the shade where stems are etiolated. Spines are most common on shrubs or small trees which often form dense thicket patches in open sunlit meadows.

Structural defences, unlike chemical defences, vary with different feeding apparatus and behaviour of the herbivore. For example, the efficacy of spines of different length and density can vary allometrically with the browser's body size. Black rhino (800-1000kg), in one study, bit off stems up to 16.3 mm diameter whereas blue duiker (3-5kg) only removed stems up to 2.3 mm diameter (Wilson & Kerley 2003). Maximum stem diameter scaled as $M^{0.35}$ over six species. From a defence perspective, numerous small spines might deter shoot removal by a small-mouthed browser but be 'invisible' to a large-mouthed rhino. In contrast, very large robust spines might deter rhino but not small-mouthed browsers which could feed on the leaves between spines. Thus, the structural defence system of a plant might vary depending on the herbivores most likely to feed on the plant and/or most impacting on its fitness.

Spines are a very common structural defence but other types of structural defence exist varying with the mode of feeding. A most remarkable defence, for those familiar with spiny plants, occurs on islands which were home to giant browsing birds; moas in New Zealand, and elephant birds on Madagascar (Greenwood & Atkinson 1977; Bond et al. 2004; Bond & Silander 2007). Instead of thick branches bearing large spines, plants defended against bird browsers have very thin, wiry branches with high tensile strength and, often, right-angled branching. The thin stems are difficult to clamp with a beak, do not break when tugged and extend like a Jack-in-the-Box when tugged, reducing the force applied by the bird (Figure 4.13). Moas and elephant birds are

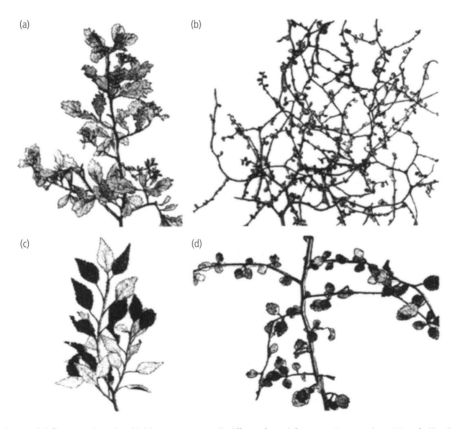

Figure 4.13 Structural defences against giant bird browsers are utterly different from defences against ungulates. Moas fed by clamping and tugging leaves and shoots. The thin twigs of juvenile stems (b, d) within moa browse height are difficult to clamp, have high tensile strength and are very flexible. Adult stems (a,c), above browse height, have thicker stems that break readily on tugging. (a,b) *Pennantia corymbosa*, (c,d) *Plagianthus regius*. From Bond et al. (2004), *Oikos*.

extinct but Bond et al. (2004) could observe feeding behaviour of ostriches and emus, their ratite relatives. Reconstructing open brown world vegetation where the consumers are extinct is not straightforward. Because plant structural defences exploit the attributes and behaviour of the browsers, it is hard to recognize structural defences for extinct creatures with no analogues to living creatures. How did plants defend themselves against giant ground sloths, for example? Nevertheless, careful observation of extant relatives of extinct animals can be very revealing as to the nature and functioning of anachronistic defences. High concentrations of anachronistic structurally defended plants can point to the ancient feeding grounds of their extinct partners.

Structural defences are best expressed in juvenile stages, just as attributes for surviving fires are strongly expressed in savanna saplings. In striking contrast to the pole-like architecture of trees in frequently burnt savannas, saplings in brown world savannas form cage-like architectures. The dense ramification protects the interior stems from loss to browsers (Figure 4.2; Archibald & Bond 2003; Charles-Dominique et al. 2017b). Cage architectures are very costly in forests where they would slow growth rates, prolonging life in the shade, and are rare in frequently burnt savannas where they prolong life in the flame zone.

4.10.2 Herbaceous plants

Grasses are also good indicators of black versus brown world. C_4 grasses are generally absent from the shaded interior of green world forests. C_3 grasses are more tolerant of shade but forest grasses seldom produce sufficient biomass to burn. Bamboos are an exception. Variation in shade tolerance of bamboos would repay further study as some bamboos seem to be characteristic of open habitats and others tolerate more shade.

Different grasses with different traits occur in tall, frequently burnt savannas versus short, heavily grazed areas (Hempson et al. 2015b). The latter, 'grazing lawns', are characterized by grazing-tolerant grasses that can spread under heavy grazing by clonal expansion from stolons or rhizomes. Bunch grasses rely on seeds to spread and their populations may fail to grow where inflorescences

are consumed under heavy grazing pressure. Grazing lawn grasses are often dismissed as indicators of 'overgrazing' or 'degradation'. However, they are a distinctive feature of semi-arid to mesic savannas in Africa. They are very productive and attract a suite of mammals capable of feeding on the leafy short grasses, including white rhinoceros, wildebeest, warthog, hartebeest, and impala. These short, heavily grazed swards lack enough fuel to burn and form grazer-created firebreaks. If animals move off the lawns, or, in the case of the extinct grazing lands of the world, the grazers go extinct, the lawns are rapidly suppressed by tall grasslands which burn. For more on grazing lawns as the most extreme example of tropical grazer-dominated open ecosystems, see McNaughton 1984; Waldram et al. 2008; Hempson et al. 2015b).

Analogous grazing-tolerant grasses occur in temperate and cooler regions of the northern hemisphere (Coughenour 1985; Linder et al. 2018). As in the tropics, they are associated with structurally defended woody plants fed on by mammals with a mixed grass/tree diet. Mixed feeders are herbivores which build large numbers by grazing grass but retain high quality food intake by switching to browse when grass quality declines in the dormant season. Heavy browse pressure, especially on woody seedlings and small saplings, helps maintain open habitat.

4.11 Open habitat faunas

'That large animals require a luxuriant vegetation has been a general assumption which has passed from one work to another; but I do not hesitate to say that it is completely false.' Charles Darwin, *Voyage of the Beagle,* Chapter 5.

There are common misconceptions about how animal assemblages change from closed to open ecosystems. Darwin, in *Voyage of the Beagle,* discussed the kind of ecosystem that supported the huge extinct animals that he excavated from the floodplains of the River Plate in Argentina. He considered whether it was a rich, luxuriant vegetation to support such numbers of these enormous beasts, such as in the jungles of India. However, he noted from correspondents that enormous herds of animals occurred in the 'poor and scanty vegetation' in the interior parts of southern Africa. So perhaps the extinct mammals lived in vegetation not unlike the existing one of an open grassland and dry woodland. He was right. Grazing lawns are the terrestrial analogue of phytoplankton with very low phytomass supporting large mammal biomass.

Here I consider how the fauna of closed forests differs from that of open ecosystems. How do animals differ in diversity, habitat endemism, and in functional traits? Which vertebrate species not only exploit open habitat but also help create and maintain their open ecosystem? Answers to these questions help inform evaluations of the conservation value of open ecosystems and what would be lost if they were converted to other land uses or planted to trees.

4.11.1 Species richness

Measures of species diversity vary with the scale of analysis. Alpha diversity is local diversity measured within a community, for example by sweep nets for flying insects, or pitfall traps for ground-dwelling creatures. If you sample birds in a patch of forest and then a patch of grassland, you would expect to find more species in the forest than the grassland. This is because structurally complex habitats, such as forests, provide more niches for feeding and breeding and should therefore support more animal species than structurally simple habitats such as grasslands. Habitat diversity, as measured by the vertical distribution of foliage (MacArthur & MacArthur 1961), has repeatedly shown this pattern in different bird communities around the world. The positive relationship between diversity and vegetation structural diversity has proved quite general, not only for birds but also flying insects, small mammals, and herpetofauna (Cody 1981). Interestingly, many studies found that vegetation structure was more important than plant species composition in determining the relationship. Any process that reduces structural diversity, such as frequent fires or heavy grazing pressure, should therefore reduce the number of species at a sampling point. Based on these arguments, we would expect the alpha diversity of fauna to generally be lower in open ecosystems than in closed forests.

Species richness is only one metric for measuring biodiversity. We are also interested in whether there

are different kinds of species as you move from one habitat to the next. Species turnover across different habitats is beta diversity, also referred to as habitat heterogeneity. To measure beta diversity, you not only count species but you also need to identify them to determine whether composition changes. Ordination techniques are commonly used. In a forest/grassland mosaic, for example, some species might occupy forest niches while others exploit savanna resources. A forest/grassland mosaic landscape might then support higher diversity than either forest or savanna on its own. For this to be the case, each habitat would need to support species endemic to that habitat. If the open habitat fauna is merely a subset of the forest community, as you might expect in a recently deforested area, then turnover will be low. But if the grassland patch has endemic species, absent from forests, then the forest/grassland mosaic will have more species than either habitat on its own. It then becomes interesting to explore the origins of open versus closed habitat faunas, since we are apparently dealing with a system stable enough for a habitat-specific fauna to evolve, and not merely an ephemeral successional stage.

Open versus closed-forest habitats represent extremes of habitat heterogeneity. But species respond to much more subtle variations in vegetation structure. Consumers alter vegetation structure and there are many studies of the impacts of different fire regimes on faunal diversity by altering vegetation structure. Similarly, mammal herbivory (and trampling) alters vegetation structure, changing the height of the grass sward, for example, with strong effects on insect diversity (e.g. Kruess & Tscharntke 2002). These studies are important in guiding fire and grazer management where biodiversity is a concern (e.g. Archibald et al. 2005; Fuhlendorf et al. 2006, 2009; Bowman et al. 2016; Donaldson et al. 2018). Manipulating fire regimes to promote biodiversity has been a major focus of research in Australia. The continent has suffered extremely high mammal extinction rates losing 28 terrestrial species since 1788. Small marsupials are particularly vulnerable, but several bird species are also threatened (Woinarski et al. 2015). Loss of heterogeneity of habitat due to changed fire regimes is one hypothesis for the high extinction rates. Aboriginal burning created a fine patchwork of fires contrasting with far more extensive fires after European settlement. The loss of habitat heterogeneity created by different post-burn ages was thought to have led to losses of vertebrate species. Recent studies suggest the main culprits causing endemic mammal extinctions are introduced predators, particularly domestic cats and foxes that have gone feral (Woinarski et al. 2015). The dingo, introduced into Australia some 4000 years ago, first initiated the wave of predator-related extinctions. The homogenization of habitat may contribute to the extinction cascade by altering resource availability and influencing hunting success of the alien predators, so that appropriate fire management for heterogeneity remains a major goal in conservation areas.

Habitat heterogeneity has become a major concern in conservation biology. While a variety of habitats, including forest/OE mosaics, is thought to be desirable for protected areas, an important further consideration is the area of each habitat patch making up the landscape mix. Habitat specialists require a minimum patch area to ensure occupancy (Diamond 1975). If, say, an open grassland patch is too small, some habitat specialists will avoid it, regardless of whether the vegetation structure is suitable for them. The scale of patch size required by open habitat specialists is poorly known. From the limited data available, species that prefer more open (tree-less) patches require larger minimum areas than closed habitat specialists. In a study of bird species in a South African savanna, Krook et al. (2007) focussed on bird species specializing on grazing lawns. Grazing lawns, maintained by white rhino and a diverse antelope fauna, vary greatly in size but have very similar sward structure. Though they have far fewer bird species than tall grass areas, lawns are the only habitat used by several specialist species. But, despite very similar sward structure, lawn specialist birds were absent on smaller lawns. Different lawn specialists had different minimum patch area requirements (Figure 4.14). Thus, it is not enough to manage for habitat heterogeneity. You also need to be aware of minimum patch sizes for habitat specialists.

4.11.2 Forest/open ecosystem mosaics

What is the contribution of OEs and forests to total biodiversity in habitat mosaics? This is often a key

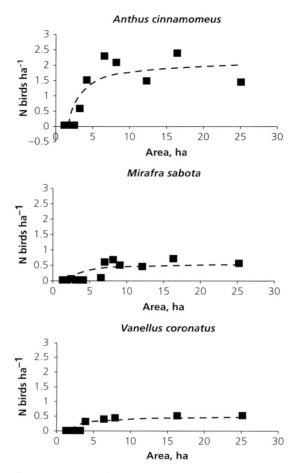

Figure 4.14 Density of birds on grazing lawns versus lawn area. Species only occupy preferred habitat patches if they exceed a minimum threshold area. Data from Krook et al. (2007).

question for conservation managers. What would be the impacts of losing open habitats, for example through long-term fire suppression to protect forests? Should managers protect the forests by suppressing fires? Or would it be better to maintain the grasslands by prescribed burning at the risk of damaging the forests? Here it is not the number of species in closed versus open habitats that counts but whether the species composition differs (beta diversity). At the landscape scale, or that of a protected area, forest/OE mosaics have occasionally been sampled. Since methods for sampling are partly habitat dependent, and biologists tend to specialize in either forest or open vegetation, there are rather fewer comparative studies then one might expect (Parr et al. 2012).

At larger biome scales, there are few generalizations on the distribution of diversity. For example, in American prairies (Central Grasslands), most plant, insect, and bird species are apparently not prairie endemics but also occur in forests and woodlands. Anderson (2006) noted that only 11.6 per cent of mammal species are true grassland species while only 5.3 per cent of the birds are 'true' prairie species. Even the grasses are thought to have evolved in forest openings, in the deserts of the south-west, or in montane meadows, and not in the prairies (see Anderson 2006 for sources). The lack of endemism in the prairie biota is a striking contrast to the savannas and grasslands in the south of the USA which are now recognized as exceptionally rich in endemic plants and animals (Noss et al. 2015).

4.11.3 Functional traits

As for plants, one would expect functional traits of animals to change from forest to open ecosystems. Resources are distributed differently in space and time, predator avoidance requires different strategies, and microclimates will be strongly divergent. Among invertebrates, ant communities show striking compositional and functional differences between open and closed habitat (Parr et al. 2012; Andersen et al. 2012; Andersen 2018; Vasconcelos et al. 2017). These differences are so profound that Andersen (2018) argued that the divide between closed and open habitat ant faunas is the basis for understanding ant community responses to disturbance. He proposed five principles for how ants respond to disturbance (fire in his examples) based on whether they are open habitat or closed habitat species (Table 4.4).

Andersen (2018) notes that 'many faunal impacts of habitat disturbance are fundamentally related to habitat openness, the effects of disturbance on it, and the functional composition of species in relation to it.' The fauna, it seems, like the flora, is divided into sun-loving, shade-intolerant taxa versus shade-tolerant forest lineages. The origins of the fauna can help explain likely responses. Thus Australia has an extraordinarily diverse ant fauna in pyrophilic savannas. The fauna has its origins in the arid interior of the continent and is pre-

Table 4.4 Five key principles for predicting ant community responses to disturbance (Andersen 2018). The principles may be more generally useful for other vertebrate and invertebrate groups.

1	The major effects of habitat disturbance on ants are typically indirect, through changes in habitat structure, microclimate, resource availability, and competitive interactions
2	Habitat openness is a key driver of variation in ant communities
3	Ant species responses to disturbance are largely determined by their responses to habitat openness
4	The same disturbance will have different effects on ants in different habitats, because of different impacts on habitat openness
5	Ant community responses to the same disturbance will vary according to ant functional composition and biogeographic history in relation to habitat openness

adapted to the open sunny conditions created by savanna burning. The forest ant fauna, in contrast, has its origins in the rainforest of East Asia and responds negatively to disturbances that open up the vegetation. In contrast to Australia, the savannas of Brazil evolved from forest tree species as did the ant fauna. The ant communities respond quite differently to frequent fires reflecting their forest origins (Vasconcelos et al. 2017; Andersen 2018). Andersen's principles of faunal responses to 'disturbance' need testing for a wider range of taxa. They imply that the OE–closed forest divide is just as important for fauna as it is for the flora.

4.11.4 Mammals

How do mammal traits vary from closed to open habitat? In forest/grassland mosaics, there will be obvious differences in meso- and megaherbivore diet with browsers in the forest and grazers in the grasslands. Along with diet, there are also differences in sociality, predator avoidance behaviour, speed, endurance, and migration (Figure 4.15). Functional differences in large mammal consumers are particularly interesting when they feed back to the potential of these herbivores to create and maintain open ecosystems. Following HSS, the most influential consumers will be those that escape predator control. Large body size, megaherbivores >1000 kg, are prime candidates (Owen-Smith 1988). These species are too big and too fierce for predators to control their population size, so they would have grown until constrained only by

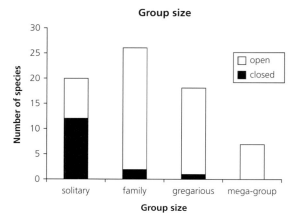

Figure 4.15 Open ecosystems provide greater opportunities for social living in African herbivorous mammals. Group sizes: solitary =1 or 2 individuals, family = 5–15, gregarious = 30–150, mega-groups > 500. Data from Hempson et al. (2015a).

resources. Consequently, they should have had large impacts on the vegetation—the keystone megaherbivore hypothesis of Norman Owen-Smith (1988). But there are other routes to avoiding predators, particularly in open habitat. Large herd size, in open habitats where social communication is facilitated, also provide protection from predation (but not parasites or pathogens) allowing large animal densities to develop. OEs, especially grassy ones, vary in seasonal food quantity and quality. But, unlike forests, OEs also facilitate untrammelled movement, allowing seasonal migration to exploit spatially distributed resources. Migrant herds escape predators in space by moving on. Predators such as those in the cat family have slow developing young and cannot follow a moving herd while rearing young. The effect of tens to hundreds of thousands of mouths feeding and hooves trampling as herds migrate is a powerful contributor to creating 'brown worlds'. Examples come from caribou (reindeer) in boreal regions, saiga antelope in the forest/steppe mosaics of Central Asia, pronghorn antelope and bison in North America, and wildebeest in Africa.

Among resident ungulates, mixed feeders seem particularly effective at limiting tree populations by feeding on seedlings and saplings. Herds can be built up by feeding on the abundant grass and then exert heavy browsing pressure when the animals turn to browse. Mixed feeding species are few in Africa, where most ungulates are specialized to browse or graze (Cerling et al. 2003). They include the widespread impala, implicated as the key agent limiting tree populations in several studies (Prins & van der Jeugd 1993; O'Kane et al. 2012; Staver & Bond 2014). Prominent mixed feeders elsewhere include goats and sheep, white-tailed deer, red deer, pronghorns, and reindeer.

Megaherbivores have dwindled and are now seriously threatened with extinction in most regions. Large size confers the ability to consume food of poor quality avoided by meso-herbivores. An interesting example is hippo in tall, humid West African savannas. The tall grasses are of poor quality, but hippos have been able to convert the tall grasslands to highly palatable short-grass grazing lawns by regularly cropping in the grazing zone accessible to their daytime wallows (Verweij et al. 2006). Grazing megaherbivores, like the hippo and white rhino, act as ecosystem engineers by creating short-grass swards that prevent fires from spreading. Their removal can switch a system from brown to black world as short grasses are replaced by tall, flammable grass swards (Waldram et al. 2008; Cromsigt & te Beest 2014). Elephants are ecosystem engineers where they fell trees or kill trees by de-barking followed by fire. Terborgh et al. (2016a,b) have argued that differences in diversity and composition of African and neotropical forests may be attributable to elephant browsing of saplings in African forests. Whether megaherbivores are, indeed, a class apart in terms of their effects on systems is, in my view, still to be tested.

4.12 Summary

Given the rich nature of open ecosystems, it seems inconceivable that there is still a widespread belief that open ecosystems, and especially the tropical grassy biomes, are of anthropogenic origin. In the next chapter, I explore the origin of the biota, based on fossils and phylogenetic evidence, which provides further insights into the antiquity of open ecosystems.

Understanding functional attributes characteristic of open vegetation can point to the most appropriate consumers needed to maintain the system. Otherwise inexplicable switches from a set of dominant trees to a quite different one become explicable from analysing sapling resistance to fire, herbivory, or shade-casting ability.

CHAPTER 5

The origins of closed and open ecosystems

Fossils and phylogenies

5.1 Introduction

Just as satellites have expanded the spatial dimension of ecology from Darwin's 'tangled bank' to the whole Earth, new technologies have also opened up the temporal dimension. New tools for interpreting fossils have proved particularly useful in exploring the domain of uncertain ecosystems. They include the development of stable isotopes as tracers of diverse factors including the evolution of C_4 grass-dominated ecosystems in the tropics. There have been remarkable developments in understanding palaeo-fires due to advances in recognizing fire-altered material, including charcoal, in the fossil record (Scott 2010). The discovery of 'mesofossils' in the late 1970s, has opened up new worlds of analysis because of the very fine preservation of organs such as flowers and fruits as charcoal (Friis et al. 2011). There have also been developments in recognizing open ecosystems. Hypsodonty, high crowned teeth common in grazing mammals, was interpreted as an adaptation to silica-rich grass food. Instead, it has been associated with dust and grit characteristic of open, non-forested ecosystems (Dunn et al. 2015). Leaf properties of some Australian plants are also providing novel insights into the presence of open ecosystems where older interpretations were of closed forests (Jordan et al. 2014). The melding of ecophysiological discoveries with paleoecology has resulted in increasingly sophisticated interpretation of fossils and the functional attributes of ancient organisms. For example, recognition of the importance of dense leaf venation for reducing resistance to the transport of water to the leaf, and the implications for photosynthetic rates, was shown to be a step change in early (Cretaceous) angiosperms leading to their higher photosynthetic rates and contributing to their ecological success relative to older gymnosperms and pteridophyte lineages (Boyce et al. 2009; Brodribb & Feild 2010).

5.2 Exploring the past

5.2.1 Fossils

Fossils are the gold standard for revealing the ecology of the past. They reveal the mix of plants present at a site and thereby provide clues as to ecosystem assemblages in ancient times. Many features of the environment can be interpreted from the geological setting and careful reconstruction of the taphonomy, the circumstances leading to fossil formation. However, there are several problems in interpreting the fossil record of open ecosystems and the processes such as fire and large vertebrate herbivory that influenced them. Fossils form in wetlands where forests are most likely to grow, even if the rest of the landscape is covered by highly flammable grasslands or shrublands. Plants are best preserved in acidic material while large vertebrate bones dissolve in those conditions and are generally found in alkaline material. Reconstructing plant–herbivore interactions is tricky when only one half of the partnership in well preserved as fossils. The sheer scarcity of fossil-bearing sites is well known and understood by palaeobiologists but can be a shock to present-day ecologists used to intense, spatially extensive sampling protocols. For example,

Open Ecosystems: ecology and evolution beyond the forest edge. William J. Bond, Oxford University Press (2019). © William J. Bond 2019.
DOI: 10.1093/oso/9780198812456.001.0001

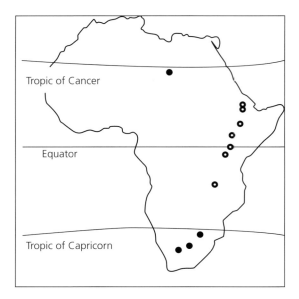

Figure 5.1 Early hominid fossil sites in Africa, 4.4 to 1 Ma. Open circles are located along the Rift Valley. Fossil sites are rare and seldom representative of the general landscape, e.g. the Rift Valley.

the heavily researched history of hominids in Africa is based on just a tiny handful of sites in East and southern Africa (Figure 5.1). It would be laughable for modern ecologists to make generalizations of such biological importance on such a small geographical sample. The (widely accepted) risk is that a given interpretation of the past is always open to radical change as a result of new discoveries.

5.2.2 Phylogenies

Fossil evidence for the past is now widely supplemented with phylogenetic inference. The rapid development of molecular methods has led to an explosion of DNA-based phylogenies filling in information that simply does not exist in the fossil record. Phylogeny reconstruction and methods for dating trees have also shown phenomenal development. While early molecular phylogenies produced dates that were wildly different from well-known fossil groups (usually much older than known fossils) there is increasing convergence between fossil and phylogenetic dates. Phylogenies reveal patterns of diversification of a lineage. It is tempting to interpret the appearance and subsequent diversification of an ecologically informative lineage as

representing major changes in landcover. The errors in making this assumption are particularly apparent in the very different timing of the origin and diversification of C_4 grass lineages (~35 Ma) as opposed to their assembly as a major biome (~7Ma) more than 20 Ma later (Bouchenak-Khelladi et al. 2014). The early dates of diversification might indicate rapid speciation as these grasses colonized small fragmented tree-less patches in a generally forested landscape. A useful approach to dating the origins of novel ecosystems is to explore the dated phylogenies of plants and animals restricted to those ecosystems. For example, the emergence of fire-adapted woody clades is a more reliable marker of when fire-dependent savannas became widespread than the diversification of grasses that provide the fuel.

5.3 The early origin of forests

The key innovation resulting in mosaics of closed and open vegetation under the same physical site conditions was the evolution of trees. In addition, trees had to evolve that cast sufficient shade to exclude shade-intolerant understorey growth forms. Plants first colonized the land ~425 Ma. The first 'tree' appeared 50 Ma later in the form of giant lycopsids (club mosses), sphenopsids (horse tails), tree ferns, and the extinct progymnosperms, ancestral to seed plants (see Willis & McElwain (2014) for an accessible introduction to ancient plant life). None of these ancient tree forms would have cast significant shade. Seed plants became more common and by the Permian (~270 Ma) were dominant in the world flora. The conifers, some of which might have cast dense shade, appeared about 250 Ma ago, becoming dominant in fossil floras by the Jurassic (200–145 Ma).

The vegetation produced by these ancient growth forms is too remote from present-day closed forests to interpret in terms of alternative states of open and closed ecosystems. What we do know, is that fires were common, particularly in the high oxygen world of the Carboniferous (359–299 Ma) and the Permian (299–252 Ma). Given the diversity of 'fuels' (flammable vegetation), and the lack of modern analogues, Scott and Glasspool used 'inertinite' in coals as a measure of ancient fire

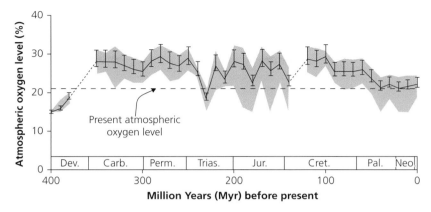

Figure 5.2 Changes in atmospheric oxygen as interpreted from inertinite content of coal for the 400 million years that fires have been burning terrestrial plant life. (From Glasspool & Scott (2010), *Nature Geoscience*).

activity (Glasspool & Scott 2010). The charcoal content of mires, wetlands, has relatively similar fuels through the ages, and can be used to interpret changing fire activity over geological time. Mires produced coal and coals have formed all over the world for much of the history of terrestrial plant life. Their studies indicated varying fire activity in different periods which they interpret as caused by varying palaeo-atmospheric oxygen concentrations (Figure 5.2). According to their analyses, fire has been important in shaping vegetation for hundreds of millions of years. In the fiery landscapes of the past, tall forests would have been restricted to topographic positions that restrict fire spread, such as floodplains and wetland margins (Falcon-Lang 2000). The evolution of new features of the plants may also have contributed to fire activity by altering fuel properties. Belcher et al. (2010) observed an abrupt five-fold increase in fossil charcoal in the earliest Jurassic of east Greenland. They noted a parallel shift in leaf type across the Triassic/Jurassic boundary from broad to fine leaves that experimental studies showed would produce more frequent litter-fuelled surface fires. Litter properties in modern mixed conifer forests in the western USA have been shown to alter fires, and feed back on plant community structure in contemporary ecosystems (Schwilk & Caprio 2011). Thus changes in leaf properties may have had feedbacks to the fire regime causing changes in the dominant species of ancient forests.

5.4 The Cretaceous and the emergence of the flowering plants

5.4.1 The attributes of early angiosperms

Flowering plants (angiosperms) first appear in the fossil record in the Cretaceous from ~139 Ma. From their early beginnings in 'damp, dark, disturbed' shaded understoreys, angiosperms emerged into open disturbed habitat by the mid Cretaceous (~90 Ma ago) (Feild & Arens 2005). By the late Cretaceous, they dominated tropical latitudes of the palaeo-world and were present, as minor components, in still-dominant gymnosperm forests at higher latitudes (Crane & Lidgard 1989). Today, uncertain ecosystems, both closed and open, are largely angiosperm-dominated so that the origin of contemporary closed forests and alternative open vegetation states must be sought in the revolution leading to the widespread dominance of flowering plants.

By the mid Cretaceous (~90 Ma ago), flowering plants had acquired many 'weedy' characteristics typical of open ecosystems (Wing & Boucher 1998; Friis et al. 2011). Reproductive innovations were well suited to disturbed high light environments. Compared to gymnosperms, flowers allowed rapid maturation and rapid seed production while angiosperm seeds were small and thought to have evolved new forms of dormancy (see Bond & J. Midgley 2012 for a review). Large seeds typical of closed forests only appeared after the Cretaceous when the angiosperms had already replaced ancient ecosystems over much of the earth (Tiffney 1984). Vegetative innovations, especially vessels, and high leaf vein densities, allowed the evolution of thinner leaves with higher photosynthetic rates than the ancient plant lineages (Boyce et al. 2009). Trees, as represented in the fossil record by large stem diameters, were very scarce for much of the Cretaceous, surprisingly so given that fossils are most likely to develop adjacent to wetlands. Fossil trees do, however, appear towards the end of the Cretaceous.

5.4.2 Dinosaurs and the spread of the angiosperms in the Cretaceous

The weedy nature of Cretaceous angiosperms is intriguing. How do 'weeds' take over the world? Bakker (1978) was the first to suggest that consumers, dinosaurs, were a major agent promoting flowering plants. He argued that the evolution of short-necked (ornithischian) dinosaurs, such as rhino-like *Triceratops*, placed heavy browsing pressure on small regenerating plants in contrast to long-necked (sauropod) dinosaurs browsing on established trees (Figure 5.3). Heavy browsing would have selected

Figure 5.3 The dinosaur hypothesis for angiosperm evolution (Bakker 1978). Enormous long-necked sauropod dinosaurs (right) were common in the Jurassic feeding on foliage of trees. In the Cretaceous, these were replaced by ornithischian dinosaurs (middle) placing heavy browse pressure on low-growing plants. 'Weedy' angiosperms would have been favoured under high herbivore pressure because of rapid growth and reproductive rates. The elephant on the left gives scale. Critics note that though the rise of angiosperms was worldwide, the switch in dinosaurs and their feeding height was not.

for rapid growth and rapid reproduction of juvenile plants analogous to the grazing lawns of today. Flowering plants have these attributes so, he argues, dinosaurs 'invented flowers'. One way of testing the hypothesis is to determine whether angiosperms spread in regions which lacked the short-necked dinosaurs and therefore lacked the selection pressure that should promote angiosperms. Barrett and Willis (2001) found no support for Bakker's hypothesis using this test since angiosperms evolved regardless of the composition of local dinosaur faunas. Nevertheless, Bakker is among the few to have suggested coevolved links between vertebrate herbivores and the plants they fed on. It seems highly likely that dinosaurs helped shape Cretaceous vegetation, not only through their feeding activity but also by their massive trampling effect (dinoturbation) as they moved through Mesozoic ecosystems.

5.4.3 Fire and the spread of angiosperms in the Cretaceous

Bond and Scott (2010) explored the other major consumer, fire, as a potential agent promoting the Cretaceous spread of angiosperms. They suggested that angiosperms created a novel fire cycle, analogous to the grass fire cycle (D'Antonio & Vitousek 1992) thereby promoting open habitats that favour 'weedy' angiosperms. These would be able to rapidly colonize post-burn gaps due to their reproductive innovations and rapidly fill gaps with flammable biomass due to their vegetative innovations. The effect of angiosperms would thus be to promote more frequent fires which would promote greater angiosperm dominance which would promote more fire, etc. The only ancient growth form likely to have created similar fire cycles are the ferns. Fossil evidence indicates that 'fern savannas' were present in the early Cretaceous, with abundant evidence of fire (Collinson et al. 2000). Ferns, especially bracken (*Pteridium* spp.) continue to produce highly flammable fuel in contemporary ecosystems. However, angiosperms, with the advantages of seeds rather than spores for dispersal and establishment, could spread into diverse, drier habitats.

There is considerable fossil evidence supporting the idea that fires helped promote the spread of angiosperms. Mesofossils are small (3–4mm), often beautifully preserved fossils first recognized by Friis in Cretaceous sediments in Sweden and subsequently found to be widespread in Cretaceous sediments. Mesofossils have revolutionized our understanding of early angiosperm evolution because of the beautifully preserved fossil flowers, fruits, leaves, and seeds (Schonenberger 2005; Friis et al. 2011). The fine detail preserved in the fossils is because they are charcoalified (Figure 5.4). Comparison with modern fires indicates that this kind of preservation is typical of low intensity shrubland fires. Thus the fossil evidence supports the widespread occurrence of fires in the Cretaceous consistent with estimates of high atmospheric oxygen at the time. Frequent fires, argued Bond and Scott, eliminated gymnosperms unable to persist in the new fire regime and promoted the spread of angiosperms in those environments warm enough and wet enough to allow for their potential for rapid growth and reproduction.

Recently, Belcher and Hudspith (2017) measured fuel properties of analogues of plants occupying Cretaceous understoreys and used the results to model how angiosperms might have changed Cretaceous fire properties. Their results suggest that the biggest impact was of angiosperm shrubs on fireline intensity. The effect would have been higher scorching of gymnosperm tree canopies and more frequent crown fires. Consequently, more intense fires would have opened up tree canopies, promoting light-loving Cretaceous angiosperms, creating fuels for more intense fires, and so on; a positive feedback for angiosperm spread.

5.4.4 Phylogenetic evidence for fire-adaptive traits in the Cretaceous

The argument for a change in fire regimes in the Cretaceous promoting the spread of open, light-demanding angiosperm-dominated communities

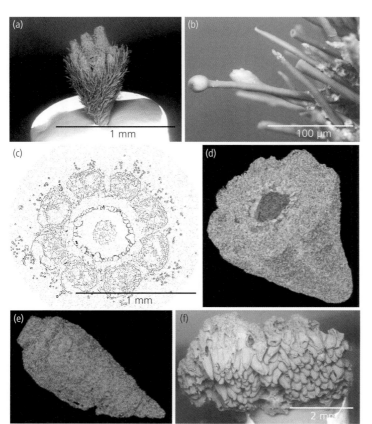

Figure 5.4 Charcoalified fossils preserve remarkably fine details as in these 335 Million-year-old seed fern fossils. Scanning electron microscopy of: a) a 1mm long seed fern ovule with b) detail of glandular hairs c) internal cross section of the ovule non-destructively imaged by X-ray tomography d) reconstructed 3-D ovule from multiple cross-sections that can be digitally stripped away showing e) the internal megaspore. Such techniques have been used to study f) pollen organs from the same deposit. Reproduced with kind permission of Andrew Scott from Burning Planet, Oxford University Press, 2018.

has been tested with molecular phylogenies, especially by Byron Lamont, Tianhua He, and colleagues in a number of lineages. He et al. (2012) used a phylogeny of *Pinus* to determine when thick bark, resistant to surface fires, and serotiny (seed storage in the canopy) characteristic of more intense crown fire regimes, had evolved. Both appeared ~89 Ma ago in the mid Cretaceous, consistent with the fossil record

of increasing angiosperm-fuelled fires. For early angiosperm fire adaptations, He et al. (2011) explored the origin of *Banksia* (Proteaceae), a common component of flammable heathlands in Australia. Here too they found a mid-Cretaceous origin for the lineage, indicating far more ancient origins of flammable heathlands than the late Miocene, nearly 80 Ma later, which had previously been proposed (Figure 5.5;

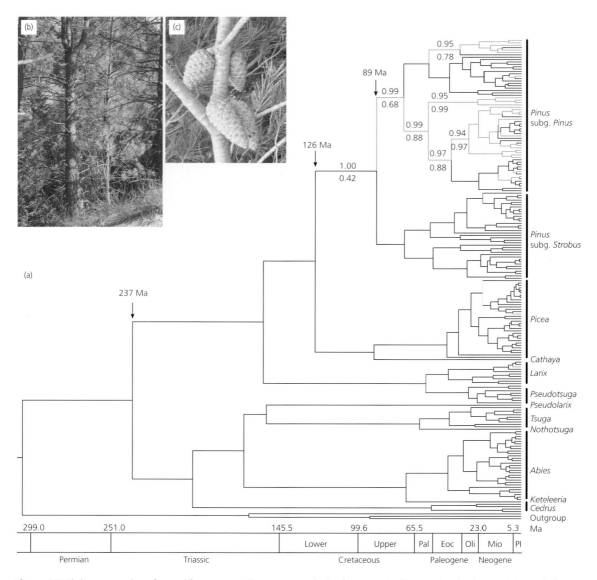

Figure 5.5 Phylogenetic evidence for novel fire regimes in the Cretaceous selecting for serotiny in *Pinus*. The dated molecular phylogeny indicates the evolution of closed cones, opening after crown fires, at ~89 Ma, in the mid-Cretaceous. (From He et al. (2012), *New Phytologist*.) (See Plate 10.)

Plate 10). Their phylogenetic hypothesis has recently been tested by a study of fossil material in late Cretaceous (Campanian-Maastrichtian) materials in central Australia (Carpenter et al. 2015). Fossil pollen and leaves of Proteaceae indicated a nutrient-limited, open, sclerophyllous vegetation with abundant charcoal as evidence of fire, and not the expected forests. This study is perhaps the first fossil test of a phylogenetically derived hypothesis. Since far fewer Cretaceous fossil sites have been examined for evidence of fire in the southern hemisphere (Brown et al. 2012), the study was particularly valuable in adding to the fossil evidence for Cretaceous fires and open non-forested ecosystems (see also Carpenter et al. 2017 for further fossil evidence of Cretaceous fires in Australia).

5.5 The first angiosperm-dominated forests

There is a great deal of uncertainty as to when angiosperm-dominated forests first appeared. Phylogenetic studies placed the emergence of tropical forest lineages, and the existence of tropical forest habitat, as mid-Cretaceous (Davis et al. 2005). The phylogenetic analyses assume that the ancient ancestors of modern tropical forest lineages were trees, not shrubs. Following this assumption, tropical forest habitat would first have appeared about 100 Ma ago. However, the first fossil evidence for tropical forest (as of 2018) is in the Paleocene, in the early Cenozoic after the dinosaurs had gone extinct (Wing et al. 2009). By the Eocene (from 55 Ma), fossil records of closed forest ecosystems are widespread around the world (Willis & McElwain 2014). Angiosperm trees dominated these forests from the tropics to the temperate zone to an extent not seen before or since. The Eocene was warm, relatively wet, and with high atmospheric CO_2 (~1000 ppm). The appearance of angiosperm trees capable of forming closed forests has been linked to a step change in leaf vein density, allowing higher photosynthetic rates, more rapid leaf expansion, and therefore the ability to cast more shade sooner than ancestral trees (Boyce et al. 2009; Feild et al. 2011). These would be necessary properties for woody plants to suppress shade-intolerant shorter plants while also producing a microclimate that inhibits

the spread of fire. Thus it seems reasonable to suggest that with the evolution of closed forests dominated by angiosperm trees, we see the first appearance of strong biotic control over the spread of fire. Before the evolution of broadleaved forests, fires would have been stopped by physical barriers such as wetlands and bare ground (Bond & J. Midgley 2012). With the appearance of angiosperm-dominated forests, we also see the birth of 'uncertain ecosystems' and the potential for alternative stable states of closed forest and open non-forest ecosystems.

5.6 The Cenozoic: the age of mammals

The Eocene was a generally warm, wet, high CO_2 world. It is of particular interest as a model for a future greenhouse world with a climate and atmosphere analogous to some projections of anthropogenic global change. Though fossil evidence indicates a predominantly forested world (Willis & McElwain 2014), flammable open ecosystems persisted, presumably in uplands and other refugia where forests could not grow. Fossil evidence for open ecosystems is rare but phylogenetic analyses indicate that lineages with fire-adaptive traits persisted throughout the Eocene. By the Oligocene (~35 Ma ago), giant mammal herbivores had evolved in many parts of the world. These included the indricotheres, the largest mammals ever to have walked the Earth and, at 20+ Mg, weighing three or four times more than modern elephants. By analogy to elephants, they must have had major impacts on forest structure merely through the mechanical effects of trampling through a forest, especially if moving at speed. Norman Owen-Smith (1987) argued that megaherbivores (animals weighing > 1000 kg), escape population regulation by predation because of their large size. If size allowed escape from predator regulation in the past, then it seems reasonable to invoke very large mammals as the potential engineers creating open ecosystems out of forests as their population grew (Y. Ayal, unpublished 2012). Indeed, estimates have been made of the effect of Pleistocene extinctions on tree cover in South American savannas by assuming that impacts on tree demography were proportional to body size of extinct mammals (Doughty et al. 2016a). The idea that massive effects are associated with massive body size so that megaherbivores of the past could have opened up Palaeogene forests deserves far more attention. Mechanical and feeding effects of large mammals have also to be weighed against recovery rates of trees depending on climate and atmospheric CO_2 at different times. In Africa, for example, as many as five species of proboscideans (elephants and their relatives) co-occurred in the same region in the Oligocene (~30 Ma) yet the vegetation was 'forest' (Sanders et al. 2010). Perhaps conditions highly favourable for rapid resprouting of damaged trees allowed the forests of the time to persist.

5.7 The origin of grasslands: another angiosperm revolution

It is hard to imagine a world without grasslands. Today, grassy biomes (savannas, steppes, tropical, subtropical, and temperate grasslands) cover some 40 per cent of the vegetated land surface. It is a most improbable success story. Grasses are easily shaded out by trees. Indeed many C_4 grasses shade themselves out if the dead litter they produce is not removed by fire or grazing. Not surprisingly, the common perception is that grasses only grow naturally where trees cannot—in arid areas, cold places, or inhospitable soils. Such habitats were, in all likelihood, the cradle of grasslands. But something happened to change all that, triggering an explosive spread of grasslands, especially C_4 grasslands, from the Miocene.

Grass phytoliths (silica bodies in leaf cuticles) have been recorded from dinosaur teeth in the Cretaceous. Leaf fossils indicate bamboos began spreading in the understorey of forests from the mid Miocene in China (Wang et al. 2013). Grass pollen has been recorded from many fossil sites in the Cenozoic (Jacobs et al. 1999). Unfortunately, grass pollen has been singularly uninformative as to the type of grass and therefore the habitats in which they grew. The best evidence for the spread of grassy biomes is not from the grasses but the animals that ate them. The evolutionary transition of equids (horses) from forest browsers to plains grazers is well known in North America where they first evolved (MacFadden 2005). Grazing equids first appeared some 15 Ma as forests began to

fragment. The cause of forest retreat has been widely interpreted as due to the onset of colder, drier climates in the Miocene. These climate changes are associated with the development of an ice cap in the Arctic. There is good supporting evidence for increased aridity in Eurasia (Tang & Ding 2013) and North America from ~10 Ma based on climate reconstructions from numerous pollen records. However, there is also evidence for a key role of consumers, especially fire but also mammal herbivory, in opening up tropical forests and replacing them with grasslands and savannas.

5.8 The origin of savannas

Our understanding of the evolutionary history of savannas was revolutionized by the discovery of the C_4 photosynthetic pathway and its diagnosis using stable isotopes of carbon. Isotopic methods could be applied to fossil organic carbon buried in ancient soils or preserved in animal bones and teeth. Isotopic analysis can reveal whether the source of the carbon is from C_3 plants or C_4 grasses. Using isotopic methods, Cerling et al. (1997) showed that C_3 dominated ecosystems were replaced by C_4 grassy biomes worldwide from about 8 Ma. The explosive spread of this new biome first began in the tropics, spreading to reach mid-latitudes (as far as ~30°S and N) by the Pleistocene. The reason for the appearance and rapid spread of the biome has been the subject of intense research interest and a fair amount of controversy over the last two decades.

5.8.1 C_4 photosynthesis and atmospheric CO_2

The key physiological innovation in C_4 plants is a CO_2 concentrating mechanism around the sites of photosynthesis. The key ecological effect is that carbon assimilation of C_4 plants is greater than that of C_3 plants at low atmospheric CO_2. The degree of advantage varies with growing season temperature, with C_4 plants doing best, and C_3 plants worst, at high temperatures. Ehleringer et al. (1997) expressed the domain of relative advantage as quantum yield on a temperature/CO_2 plane (Figure 5.6). If photosynthesis is central to ecological success, then physiological considerations predict

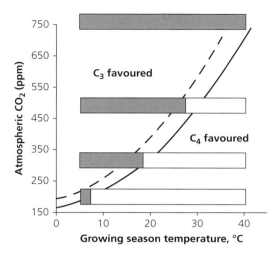

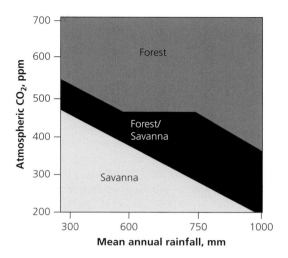

Figure 5.6 Predicted photosynthetic performance of C_3 versus C_4 photosynthesis in relation to the growing season temperature and atmospheric CO_2 based on quantum yield. The dashed and solid lines indicate the range of cross-over temperatures at which relative performance shifts. The horizontal bars indicate C_3 (grey) or C_4 (open) plants favoured. Bars highlight: 750 ppm, Eocene worldwide forest dominance; 500 ppm, tipping point for C_4 plants favoured in hot climates; 280 ppm, interglacial conditions favouring C_3 plants over wider temperature range; 200 ppm, glacial conditions favouring C_4 plants over wider temperature range. Figure based on Ehleringer et al. (1997).

Figure 5.7 Rising atmospheric CO_2 should favour trees relative to grasses leading to replacement of savannas by advancing forests. Changing distributions of forests and savannas on a rainfall, CO_2 plane with tree growth simulated by the Sheffield DGVM. See Bond et al. (2003) for simulations on which this figure is based.

that C_4 dominated grasslands would first have appeared near the Equator (high growing season temperature) when CO_2 in the atmosphere fell below ~500 ppm. Ehleringer et al. (1997) suggested that this occurred near the end of the Miocene, coinciding with the isotopic evidence for the spread of savannas. This elegant hypothesis triggered a host of studies exploring different facets of the argument. Among these was the use of diverse proxies to reconstruct historical CO_2 levels in the atmosphere. These studies showed that CO_2 levels had plunged from high levels in the Eocene (~50 Ma) to below 500 ppm, the suggested threshold, as early as the Oligocene (30 Ma) and long before the rise of the savanna biome. That the physiological arguments were credible was shown by phylogenetic studies pointing to the origin of C_4 grasses also at ~30 Ma. Though consistent with their predicted ability to perform well under low CO_2, these early dates for the emergence of C_4 grasses suggested that changing CO_2 was not the cause of the spread of savannas,

which began some 20 Ma later (Osborne 2008; Edwards et al. 2010).

Ehleringer et al. (1997) did not explain how high photosynthetic rate might lead to replacement of C_3 by C_4 plants. Effectively the argument is that differences in carbon assimilation account for the distribution of forests versus grasslands. Trade-offs in carbon allocation, to herbaceous or woody growth forms, are far more likely to account for changing abundance and distribution of trees versus grass. Recent experimental studies on tree responses to a gradient of CO_2 show profound effects on growth, fire response, plant defence allocation, and resilience to drought from low to high CO_2. The low CO_2 of glacial periods (<200 ppm) would have been a major stress factor for woody plants according to these experiments (Figure 5.7). Threshold effects below which trees do poorly and above which trees thrive have not, as yet, been explored as an alternative CO_2-linked explanation for the Late Miocene retreat of forest and advance of savannas.

5.8.2 The role of consumers: fire

C_4 grasslands and savannas are the most extensive open ecosystem in Uncertain Ecosystem Climate

Zones. How might the consumers, fire and vertebrate herbivores, be implicated in the spread of the grassy biomes? After tens of millions of years of quiescence, fires began to burn in many regions worldwide from the late Miocene (10 Ma) (reviewed by Bond 2015) (Figure 5.8). Marine charcoal records for the North Pacific collected by Herring (1985) show a surge of fire activity at all cores from this time. Similar increases in fire activity have been documented from cores in the North and South Atlantic at low to mid-latitudes. Phylogenetic analyses of plants growing in frequently burnt savannas in Brazil (Simon et al. 2009; Simon & Pennington 2012) and southern Africa (Maurin et al. 2014) place the origin of adaptations of woody plants to grass-fuelled fire regimes in the Pliocene from ~6 Ma with peak diversification from 2–4 Ma. Thus both fossil and phylogenetic information support a surge of fire activity in C_4 grassy biomes within the last few million years. Further studies on the organisms that depend on C_4 grassy ecosystems

are required to help identify when they became prominent as a major global biome.

Savannas were not the only system to experience increased fire activity. Both Herring's charcoal studies and phylogenetic analyses in woody-fuelled fire regimes indicate more fire activity and/or extension of flammable open ecosystems into previously more wooded habitats from the late Miocene (Hoffmann et al. 2015; Rundel et al. 2016, 2018). The question, then, is what triggered this apparent global increase of fire within the last 10 million years. The increase happened too early and was too widespread to be attributed to emerging hominin use of fire. There is some evidence that Neogene climates were not only becoming more arid as they became cooler but that rainfall became more seasonal ('monsoonal'). It is a common mistake to describe savanna climates as 'dry'; annual rainfall in savannas often exceeds 1000 mm, far wetter than London (~590mm pa) or even New York (~1140mm). High rainfall in the wet season is more than sufficient

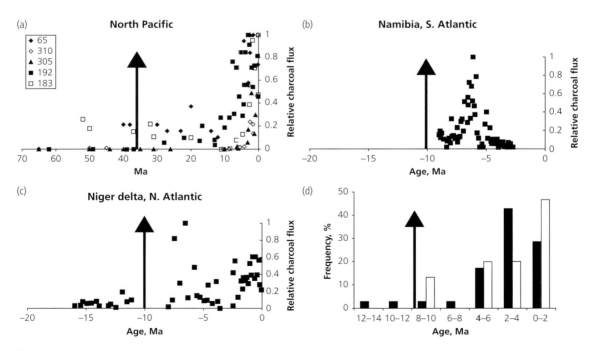

Figure 5.8 The late emergence of fire in the Cenozoic as reflected in relative charcoal fluxes recovered from marine sediments in (a) the North Pacific (from Herring 1985), (b) the North Atlantic (from Morley & Richards 1993), and (c) the south Atlantic (from Hoetzel et al. 2013). (d) shows estimated ages of fire-adapted woody clades from savannas in Africa (shaded) and Brazil (open) from Maurin et al. (2014) and Simon et al. (2009) respectively. The surge in fire activity occurred after 10 Ma (arrows).

to grow forests in savanna regions. However wet–dry seasonal cycles favour increased fire activity because high productivity in the wet season leads to the accumulation of large quantities of biomass, fuel, in the dry season. Fires propagate more readily in the dry season because the grasses dry out. Could the emergence of wet–dry climates from the Late Miocene have triggered the expanse of fire-dependent ecosystems, helping to roll back closed forests?

Keeley and Rundel (2005) elaborated on this hypothesis for the rise of the savanna biome. Greater rainfall seasonality may also account for the Neogene spread of other flammable systems, including Mediterranean-type shrublands in California and the Cape region of South Africa and in Australia (Bond 2015; Rundel et al. 2016, 2018).

5.8.3 Climate, fire, CO_2 interactions

In an elegant simulation study, Scheiter et al. (2012) explored the interactions between potential drivers of the expansion of savannas in a DGVM modelling framework. A particular strength of their study was their modelling of interactions among different factors. Using late Miocene climate reconstructions, they tested how CO_2, temperature, precipitation, fire, and the tolerance of vegetation to fire influence C_4 grassland expansion in Africa. The C_4 photosynthetic pathway is a CO_2-concentrating mechanism outperforming C_3 photosynthesis where growing seasons are warm and/or where atmospheric CO_2 concentrations are low. However low CO_2 failed to simulate the observed C_4 grass expansion because the climate supported extensive closed woody vegetation and C_4 grasses cannot tolerate shade. Fire was the essential ingredient causing simulated expansion of C_4 grasses in their model (Figure 5.9). An interesting twist was that the introduction of savanna trees, shade-intolerant but fire-tolerant, promoted the spread of savannas.

5.8.4 The role of consumers: mammals

Owen-Smith (1987) was the first to implicate mammal herbivory as a process maintaining open ecosystems. Pointing to the shift from open habitats to closed forests in North America and Europe from glacial to interglacial times, he argued that extinction of the megafauna, and particularly the megaherbivores (animals >1000 kg) allowed forest expansion in the Holocene. Evidence for the megaherbivore hypothesis remains limited (see Chapter 8). However a role for meso-browsers in the spread of savannas has recently been proposed for Africa (Charles-Dominique et al. 2016). This study explored woody plants with structural defences, stem spines, specific to mammalian browsers, as markers of high mammal activity. Using the distribution of nearly 2000 woody plant species in Africa, they showed that spiny plants were most common on nutrient-rich soils, drier climates, and savanna ecosystems. They were all but absent from closed forests. Undefended plants, in contrast, were most common in higher rainfall regions, and on leached,

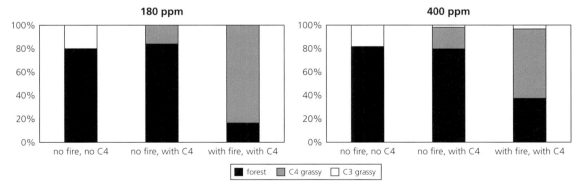

Figure 5.9 DGVM simulation of factors influencing the spread of C_4 grassy biomes in Africa in the Late Miocene. Results from two CO_2 scenarios, 180 and 400 ppm, are shown. C_4 grasses replace C_3 grasses at both CO_2 levels in both scenarios. The combination of fire and low CO_2 promotes major expansion of C_4 grassy systems at the expense of forests. Data from Scheiter et al. (2012).

nutrient-poor soils in savannas or diverse soils in forests. As regards mammal consumers, a recent continent-wide analysis of African mammals provided the necessary data to test which mammals were most closely associated with plant communities with a high proportion of spiny species. Antelope, and especially mixed feeders (consuming grass and browse), were the best candidates and most abundant in open habitats, drier climates, and on nutrient-rich soils. High rainfall savannas were less favourable with lower mammal abundance and diversity and few spiny plants. These contrasting types—eutrophic semi-arid savannas dominated by spiny plants versus dystrophic savannas in wetter, nutrient-poor areas and dominated by members of the Caesalpiniaceae and Combretaceae—are a characteristic feature of Africa (Scholes 1990). Mammals are abundant in the spiny biomes, while high fire activity is characteristic of non-spiny dystrophic savannas.

Phylogenetic analysis showed that spines had at least 55 independent origins in Africa. Remarkably, however, spines only emerged in the Miocene (~16 Ma ago) and not before, despite the long history (~50 Ma) of mammal herbivory in Africa after the dinosaurs went extinct. For much of this period, Africa was an island continent with a mammal fauna dominated by proboscideans (elephants and their relatives) and hyracoids filling the niches now occupied by antelope. Surviving hyracoids are small guinea-pig-like browsers. Their ancient ancestors had similar body shapes but some grew as large as a small rhinoceros. Major faunal changes occurred when Africa collided with Eurasia. Elephants spread out of Africa into Eurasia while horses and rhinoceros migrated into Africa, and, as the land connections grew, bovids, the lineage including the antelope of modern Africa, finally arrived on the continent (Werdelin & Sanders 2010). Antelope rapidly speciated, as did spiny plants. Lineage divergence rates of bovids and spiny plants are remarkably convergent (Figure 5.10; Plate 11) indicating that spines are particularly well-adapted to bovid modes of feeding. Given their contemporary distribution centred in savannas, the origin and diversification of spiny plants from the mid Miocene suggests a strong link to the arrival of a new class of herbivores with a novel mode of feeding. The drier savannas of Africa displaced dry forests under the pressure of browsing by this new class of mammals. Charles-Dominique et al. (2016) noted that the same phylogeny places the origin of mammal-adapted plants earlier than fire-adapted plants (Maurin et al. 2014). Thus, they suggest, African savannas first appeared ~16 Ma in response to a new class of consumers, the bovids, with fire-adapted savannas emerging several million years later.

This study is the first to point to meso-browsers rather than megaherbivores as architects of savannas. In support of their argument, it is interesting to note that parts of Africa had as many as five species of co-occurring proboscideans (elephant-like browsers) in the Oligocene in forest vegetation (Sanders et al. 2010). Despite the presence of these giant creatures, the vegetation was a closed woody system. This study raises the question of the relative importance of megabrowsers, capable of breaking down large trees, versus meso-browsers that consume the seedlings and saplings before they escape the browse zone. What is clear, however, is that both fire and mammal browsers can maintain open ecosystems in Africa and have done so for many millions of years.

5.9 Summary

Figure 5.11 summarizes the deep history of uncertain ecosystems and the history of consumers that influenced their formation. Plants first colonized the land about 450 Ma. Trees first appeared about 100 million years later. Conifers appeared in the Permian (300 Ma) and conifer forests and cycadophytes were common in the Triassic and Jurassic from 252 Ma to 145 Ma. Flowering plants first appear in the early Cretaceous from ~140 Ma. However they only become ecologically significant from the mid Cretaceous at ~90 Ma. Flowering plants assembled into new biomes starting in the paleotropics and spreading to temperate latitudes. Though present in boreal regions, these were the least altered by the angiosperm revolution. During their rapid spread, flowering plants were primarily shrublands and herbs with trees only appearing near the end of the Cretaceous. Angiosperm trees first assembled to form closed forests in the Palaeocene, from ~60 Ma. Starting in the tropics,

(a)

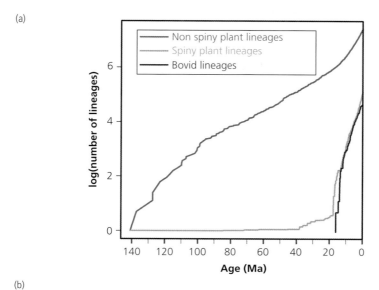

(b)

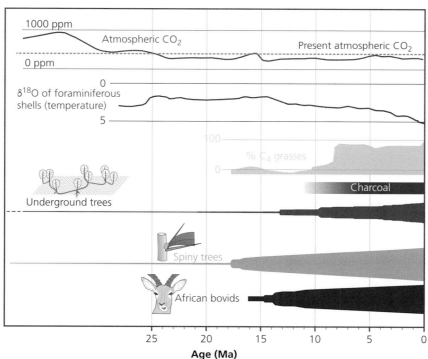

Figure 5.10 Factors influencing development of savannas in Africa. (a) Diversification of bovids (including antelope) and spiny plants. Proboscideans, including elephants, and hyracoids, dominated the herbivore assemblages in the Cenozoic until bovids entered Africa from Eurasia ~16Ma. (b) Spiny plants prefer open habitat, so that bovids may have opened up forests for savannas before fires triggered C_4 grass expansion in the Late Miocene. From Charles-Dominique et al. (2016). (See Plate 11.)

forests began to spread, reaching their peak cover worldwide in the Eocene (56–34 Ma). Broadleaved forest trees shaded out the ancestral shrubby and herbaceous angiosperm growth forms, suppressing the fires that had rolled across Cretaceous landscapes. Open flammable vegetation became very scarce in the fossil record but must have persisted in places where fossil formation is poor since shade-intolerant lineages disappear and re-appear. For example, the genus *Pinus* was widespread in the flammable Cretaceous and Palaeocene then effectively disappears from the fossil record, re-emerging only late in the Cenozoic as forests began to shrink. From their Eocene peak, forests began to fragment and grasslands, shrublands, and arid vegetation to spread as climates became cooler. Atmospheric CO_2 dipped sharply from its Eocene highs at the start of the Oligocene (~34 Ma). Open ecosystems shrank in the generally warm, wet, high CO_2 world of the Eocene.

As regards consumers, Cretaceous ecosystems had to contend with enormous dinosaurs, both long-necked sauropods and short-necked ornithischians. After the end Cretaceous extinction of the

dinosaurs, maximum body size of mammal herbivores remained below 1000 kg until the late Eocene (~40 Ma) (Saarinen et al. 2014). Thus angiosperm forests of the early Cenozoic had no megaherbivores to contend with. Forests began to fragment as climates got cooler and atmospheric CO_2 declined from the Oligocene. Gigantic mammal herbivores had evolved by this time. In Eurasia, these included *Paraceratherium*, a genus of hornless rhinoceros, which was among the largest mammals ever to have existed. They stood more than 5 m tall with a mass of 20 Mg, four or five times larger than African elephants. These giant herbivores no doubt consumed and trampled trees, and perhaps opened up forests from the Oligocene to the early Miocene. Proboscideans (elephants and their relatives) in Africa reached similar large sizes during this period and must also have smashed open forests. The history of mammals as potential initiators of major vegetation change, rather than responders to new habitat, is still poorly known (Bakker 1978; Charles-Dominique et al. 2016). Though less charismatic than the giants (Owen-Smith 1987), small-mouthed browsers appear to be very influential in limiting

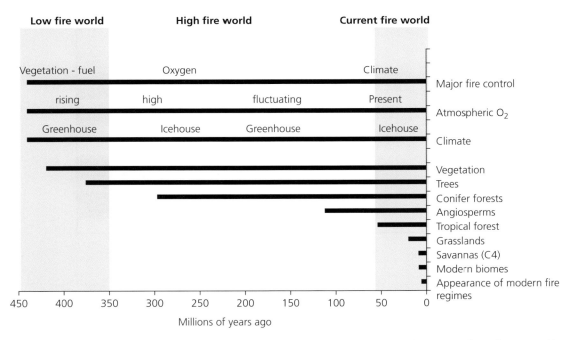

Figure 5.11 Evolution of plants and fire systems in relation to changing climate and oxygen levels. Closed angiosperm forests first appeared in the fossil record ~55 Ma. Figure redrawn from Scott et al. (2014) (Fig. 3.8, p.71, *Fire on Earth: an Introduction*, Wiley-Blackwell).

recruitment of trees and helping open up habitats (Charles-Dominique et al. 2016; see Chapter 8).

In contrast to mammal consumers, changes in fire as a consumer have been implicated in major vegetation changes at several points in earth history (Pausas & Keeley 2009; Bond and J. Midgley 2012; Belcher et al. 2013). Fire activity worldwide increased greatly from the end Miocene (~10 Ma) carving into forests and promoting the rapid spread of flammable C_4 grasslands and savannas (Keeley & Rundel 2005; Beerling & Osborne 2006; Simon et al. 2009; Maurin et al. 2014). Late Miocene increases in fire activity are also implicated in the spread of flammable Mediterranean-type shrublands. Wet winters followed by dry summers set up fuels and fire weather conducive to fire (Rundel et al. 2016, 2018). Fire in flammable woody vegetation has ancient origins, long preceding Cenozoic fires. Pyrophytic angiosperms evolved in the Cretaceous, conifers in the Permian, ferns in the Carboniferous (Belcher et al. 2013). Fires were the first major consumer of vegetation. Indeed fires were burning ancient vegetation at least 100 million years before the first vertebrate herbivores evolved. Synergies between fire and large vertebrate herbivores, and the fossil analogues, if any, of black, brown, and green world, have yet to be explored.

5.9.1 The relevance of geological history of uncertain ecosystems

This brief history of the uncertain ecosystems suggests the need for a more complex interpretation of the history of terrestrial vegetation. It is not simply a climate story overlaid on shifting continental plates with long periods allowing evolutionary innovations. The mid-Cretaceous dominance of low-growing weedy angiosperms at low to mid-latitudes cannot easily be explained by climate. Where were the forests? Nor can the origin of the savanna biome, in a similar tropical setting to those in which the angiosperms rose to prominence, be easily explained by climate alone. The historical perspective indicates the perils of purely physical explanations for contemporary vegetation patterns. Edaphic explanations for the existence of forest/savanna mosaics, for example, beg the question of what grew on these 'grassland soils' before grasses evolved. The antiquity of open, non-forested ecosystems is particularly important in recognizing their intrinsic value, given the fetish for forests in public policy and practice. But beyond contemporary issues, the past is a fascinating playground for mind-expanding discoveries and essential to understanding the ecology of our planet (Beerling 2007).

CHAPTER 6

Soils and open ecosystems

6.1 Introduction

Traditionally, climate explains major vegetation patterns within regions and worldwide (Chapter 2). 'Zonal' soils are those typical of a biome and its characteristic climate. If climate, or land-use history, does not explain the vegetation, such as in a forest/grassland mosaic, then the non-forest soils would be presumed 'azonal' (perhaps floodplain soils) and not characteristic of the biome. These bottom-up (resource-based) explanations of vegetation have been challenged for at least a century by proponents of disturbance, especially fire, as an alternative factor influencing the distribution of forests. The emergence of Alternative Stable State theory (Chapter 3) as an explanation for forest/OE mosaics has re-energized the debate over the importance of soil. A key assumption of ASS is that an ecosystem can switch to an alternative state. If soils uniquely determined which state grows where, then regime shifts to alternative states would not be possible. Thus, tests of ASS have called for evidence that both states can grow in the environments of the alternative state (Petraitis 2013).

The role of soil in determining open versus closed ecosystems is not only a theoretical issue. The widespread woody increase in many open ecosystems threatens their future. If soils are major determinants of vegetation pattern, then edaphic controls would act as a brake to major vegetation change. Among the biggest ecological changes you could get would be switches between open and closed vegetation (Chapter 4). So is there a direct deterministic soil control of open versus closed vegetation? Will some soils always be grassland soils because of edaphic barriers to tree growth? Where fires are suppressed, will forests invade the grasslands? Can we expect regime shifts, as predicted by ASS, between open and closed states depending on consumer activities and feedbacks to soil properties? The answers are likely to vary from place to place but there is not even agreement on methods for testing the competing hypotheses.

6.1.1 Hypotheses on the role of soils in shaping ecosystems

Trees, especially dense populations forming forests, require soils sufficiently deep to access the root volume necessary for water, nutrient acquisition, and anchorage. Thus, physical constraints on rooting depth might often prevent forest formation. Barriers to roots could be solid rock layers, impermeable soil horizons, or anoxic conditions in waterlogged soils. Black cracking clay soils, vertisols, swell and crack in wetting and drying cycles severing long tree roots, but not the fine fibrous roots of grasses. Such soils also tend to exclude trees. As regards chemical constraints, Lloyd et al. (2015) list three general hypotheses for nutrient constraints on forest development. First, there may be insufficient minerals to build the biomass needed for forests (Bond 2010; Silva et al. 2013). Second, a shortage of photosynthetically relevant nutrients such as nitrogen and phosphorus could lead to reduced carbon acquisition and slow tree growth. Third, growth or mortality processes not directly related to photosynthetic carbon acquisition may be affected by mineral deficiencies (e.g. Lloyd et al. 2015; Mills et al. 2016). Agricultural and forestry experience show that nutrients clearly matter for plant growth and development. There is a problem, however, in arguing that nutrient constraints alone can prevent forests from developing at a site. How do you test the soil

Open Ecosystems: ecology and evolution beyond the forest edge. William J. Bond, Oxford University Press (2019). © William J. Bond 2019.
DOI: 10.1093/oso/9780198812456.001.0001

hypotheses? Experimental nutrient addition, or experimentally altering physical constraints to root development (e.g. with explosives; San José & Fariñas 1983) only indicate short-term effects on tree recruitment and growth. But in the long term, trees can alter soil conditions in ways that overcome physical and nutrient constraints. For example, a tree seedling only exploits the surface soil layers. A fully grown tree exploits a much greater soil volume redistributing nutrients from deep to shallow layers through litter fall (e.g. Jobbágy & Jackson 2001). Feedbacks of the vegetation onto soil properties can profoundly alter soil morphology, physical properties, and especially soil chemistry. These feedbacks bedevil correlative studies for testing edaphic control.

Long-term fire suppression, or herbivore exclusion, experiments are valuable for indicating whether soil factors are more important than the consumers in preventing colonization of forest trees. They can also be used to indicate the magnitude of feedback effects on soil properties by different tree species. Paleoecological studies of vegetation change extend the timescale to centuries and millennia. In tropical grassy/forest mosaics, carbon isotopes and other proxies can show long-term stability, or rapid change, especially from open to closed forest. Such studies are very useful in demonstrating the permanence of edaphic constraints. These longer-term studies need to be better linked to the types of soil, and to the physiological mechanisms, by which plants respond to edaphic barriers to trees. With appropriate soil maps, a better conceptual understanding of how soils limit tree growth, built on a foundation of empirical studies, we will be far better placed to predict the role of soils in determining heterogeneity of vegetation.

My focus in this chapter is on the importance of soils in maintaining open ecosystems where the climate is warm enough and wet enough to grow forests. So, the key questions are:

1) Do soils alone account for OEs and for forest/OE mosaics? How would you tell?
2) Or does plant growth on different soils interact with consumers to promote or inhibit forest development?
3) What attributes of soils, physical or chemical, most inhibit tree growth?

4) Are there feedbacks between soil and vegetation that would help maintain open versus closed ecosystems?

6.2 Physical controls on forest development

Trees require greater rooting volume than grasses or shrubs for anchorage, soil moisture storage, and nutrient acquisition. In a global study, Jackson et al. (1996) found that the average maximum rooting depth for trees was 7.5m, for shrubs 4m, and for grasses 2m. Physical obstruction to tree roots is a common explanation for open patches in otherwise forested landscapes. Noss (2012) has provided a comprehensive account of edaphic grasslands in the south-eastern USA. He notes that grasslands often occur on soils that inhibit tree growth. They include soils that are shallow, alkaline, toxic, infertile, coarse-textured (sandy), or fine textured (e.g. shrink–swell clay soils). Such soils inhibit tree growth in many regions around the world. They support not only grasslands but also other light-demanding communities of shrubby or herbaceous plants. Noss provides many examples of edaphic grasslands maintained by the physical properties of soils. Their key distinction is that other factors that help maintain open ecosystems, such as fire or intense herbivory, are not required for their persistence. Examples elsewhere include the treeless grasslands of the southern Serengeti in Africa, which occur over shallow soils onto compact, alkaline, volcanically derived subsoils (Belsky 1990). Black cracking clays (vertisols) swell and shrink in wet–dry cycles tearing tree roots but with less effect on grass roots. Extensive areas of open grasslands occur on such soils in Australia (Barkly Tableland) and North America (blackland prairie soils in the Black Belt of Alabama and Mississippi). The soils are often relatively deep, and many have been cultivated to crops, especially cotton. However, the absence of trees on cracking clays is not universal. In Africa, some species of *Acacia* thrive on vertisols and one, *Acacia nilotica*, has become invasive on these soils in Australia. Furthermore, a native species, *Eucalyptus salubris*, is widespread on vertisols in Western Australia (Mills et al. 2016).

6.2.1 Waterlogging

Physical limits on tree growth are particularly common on soils that are seasonally waterlogged. The length of the 'hydroperiod' (the duration of waterlogged conditions) has pronounced effects on the type of plant community, with long hydroperiods producing wetlands and marshes (Sarmiento 1984). However, many soils, especially in the wet–dry climates of savannas, have short hydroperiods during the brief rainy season followed by long dry periods during the dry season. Such soils inhibit tree growth, producing edaphic constraints favouring open, non-forested ecosystems. They are characteristic of subdued landscapes with well-drained soils on upper slopes changing to hydromorphic soils on mid and lower slopes (Fan et al. 2017). In some tropical landscapes, forests occur on the well-drained soils and grasslands and open woodlands on the seasonally waterlogged lower slopes.

A short hydroperiod is not necessarily an absolute barrier to forests since trees colorizing the margins can draw down the water table allowing further forest tree colonization downslope. This phenomenon has been well studied in western Australia, for example, where clearing of eucalypt woodlands for wheat fields was followed by a rising water table (Peck 1978). The water table was saline, converting the soils on lower slopes to sterile sodic patches. Attempts to re-establish eucalypts have been made, using the trees as water pumps to lower the water table again. At least in principle, one can envisage alternative woodland and grassland stable states. Woodland states would originate during extended droughts with lowered water tables. Once established, trees would lower the water table by evapotranspiration. Removal of trees would result in a regime shift, raising the water table, and inhibiting woody plants in the seasonally flooded anoxic soils.

Edaphic control by seasonal waterlogging deserves more attention as a factor limiting forest distribution. In areas of subdued relief, changes in landscape drainage, for example by construction of roads, development of erosion gullies, or engineering structures for flood control, can have major consequences

Figure 6.1 Montane grasslands in Brazil with forest patches on ridge crests. The forests occur on deep red, well-drained soils and the grasslands on pale, seasonally wet soils. Fire protection for >20 years has not changed vegetation structure. Serra do Cipó National Park, Minas Gerais, Brazil.

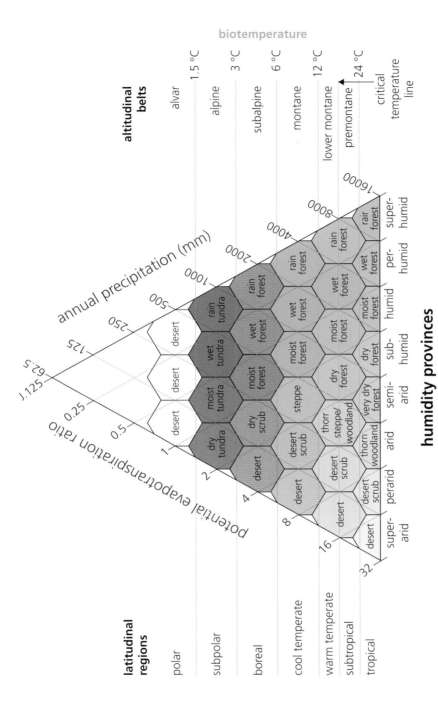

Plate 1 The Holdridge system. Each vegetation unit is uniquely associated with a climate envelope defined by two of biotemperature, precipitation, and potential evapotranspiration ratio. Biotemperature is a heat summation index calculated from average temperatures with the substitution of zero for all intervals where the temperature is below 0 °C or above 30 °C. Precipitation is the mean annual precipitation (mm). Potential evaporation is mm of water evaporated or transpired calculated as 58.93 times the biotemperature. The potential evapotranspiration ratio is the ratio of potential evaporation to annual precipitation. Values >1 imply humid climates while values <1 indicate more arid climates. Refer to Fig. 2.1 on page 14.

Plate 2 Examples of uncertain ecosystems. a) Mosaic of subtropical grassland, savanna, and forest in Gabon, Africa; b) Goias, Brazil; c) subtropical lowland forest/savanna mosaic, South Africa; d) mosaic of forest and savanna, Kakadu, Australia; e) montane savanna/forest mosaics, Drakensberg, South Africa; f) forest/ fynbos shrubland mosaic, Cape, South Africa; g) mosaic of oak parkland and mixed forest, Sonoma, California; h) Scottish moorlands with conifer plantation. Refer to Fig. 2.4 on page 17.

Plate 3 Examples of uncertain ecosystems (cont'd). (a) *Quercus robur* forest with *Festuca* grasslands, Carpathian Basin, Hungary (photo Á. Molnár). (b) *Betula pendula–Quercus robur* forest and *Stipa* steppe, Tula region, Russia (photo Yu. A. Semenishchenkov). (c) *Quercus pubescens* forest and *Stipa* steppe, Crimean Peninsula (photo Y. P. Didukh). (d) *Betula pendula* and *Festuca–Stipa* grasslands, Kostanay Region, Kazakhstan (photo Z. Bátori). (e) *Betula platyphylla* trees in *Leymus–Filifolium* grassland, Ulan Buton, Inner Mongolia, China (photo H. Liu). (f) Forest/steppe landscape with *Betula platyphylla* and *Stipa* grassland, Greater Khingan Range, China (photo H. Liu). (g) *Quercus brantii* woodland with *Bromus* grassland, Zagros Mts, Iran (photo A. Daneshi). (h) Mosaic of *Picea schrenkiana* forests and *Stipa* steppes, Xinjiang Uygur Region, China (photo H. Liu). From Erdős et al. (2018), *Applied Vegetation Science*, 21(3), 345–62. (CC by 4.0). Refer to Fig. 2.4 on page 18.

without fire

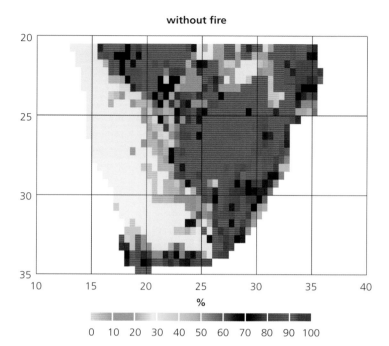

DcBI Cover with fire

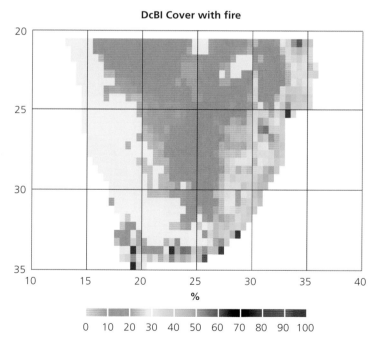

Plate 4 Tree cover in southern Africa simulated by a DGVM. The figure above indicates predicted tree cover based on the climate potential for tree growth. The figure below includes a fire module. The actual vegetation is grassland and savanna in the summer rainfall climates of the North East and fynbos shrublands in the winter rainfall climates of the South and South West. Small patches of forest occur in both climate zones consistent with vegetation simulations without fire. From Bond et al. (2003a). Refer to Fig. 2.7 on page 24.

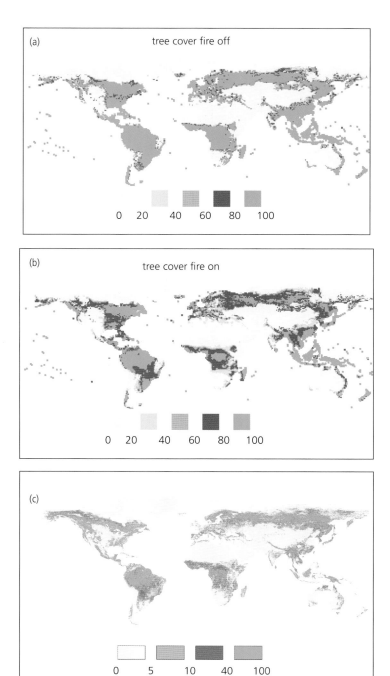

Plate 5 Global tree cover (%) (a) simulated with 'fire off' (b) simulated with 'fire on' (c) observed tree cover derived from satellite imagery in 2000. Simulated cover values are median tree cover for twentieth century simulations. Observed tree cover classes are: 40–100%, closed forest with no grass understorey; 10–40% more closed forms of savanna and other types of 'forest'; 5–10% scattered trees. The map does not discriminate between natural forests and plantations. Figure from Bond et al. (2005), *New Phytologist*, with permission. Refer to Fig. 2.8 on page 24.

Plate 6 Fire is very influential in shaping ecosystems as revealed by long-term experiments. Top: Longleaf pine savannas, Florida, (a) burned annually for 40+ years, (b) with fires suppressed over that period. Bottom: Savannas in Kruger National Park, (c) arid savanna with, left, >40 years fire suppression, right, burned every 2 years, (d) mesic savanna with, left, burned annually for >40 years, right, fire suppressed. The mesic savanna is switching to a closed thicket. Florida pictures, Winston Trollope; Kruger, W. Bond. Refer to Fig. 3.9 on page 39.

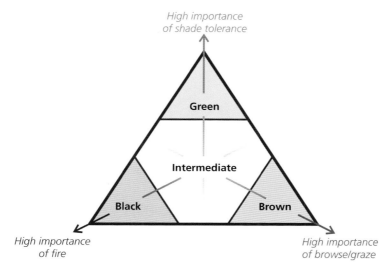

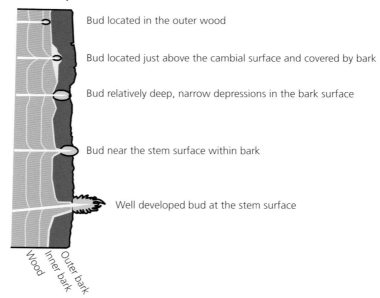

Dormant bud protection

Bud located in the outer wood

Bud located just above the cambial surface and covered by bark

Bud relatively deep, narrow depressions in the bark surface

Bud near the stem surface within bark

Well developed bud at the stem surface

Wood / Inner bark / Outer bark

Plate 7 Top: Green, black, and brown world for savanna versus forest woody plants represented as a trade-off triangle. The table lists traits grouped under those supporting growth, fire, and herbivore resistance. Fire traits are for a surface, grass-fuelled fire regime. Herbivore resistance traits are for ungulate herbivores. See Charles-Dominique et al. (2015b, 2018) for worked examples and methods used to classify trait space for tree species. Refer to Fig. 4.3 on page 50.

Bottom: Buds vary in the location in relation to bark. The relative insulation from fire injury varies from greatest at the top of the figure to least at the bottom. Figure from Charles-Dominique et al. (2015a), with permission. Refer to Fig. 4.7 on page 53.

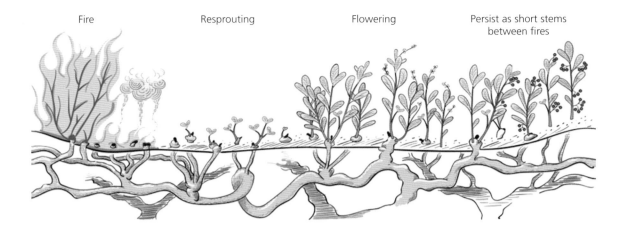

Fire   Resprouting   Flowering   Persist as short stems between fires

Plate 8 Top: 'Underground trees' (geoxylic suffrutices). Some woody species with tall tree relatives form 'underground trees', with massive below-ground branches supporting short above-ground stems that resprout rapidly after a fire (From Bond (2016), *Science*). Refer to Fig. 4.9 on page 54.

Bottom: Many members of the Eriocaulaceae, such as this population of *Paepalanthus urbanianus* Ruhland ('rocket plants'), flower en masse after fire in the cerrado of Brazil. Fire-stimulated flowering is common in many herbaceous species in crown- and surface-fire regimes in different parts of the world. Refer to Fig. 4.10 on page 55.

Plate 9 Tropical grassy biomes are rich in forb species many of which have large underground storage organs and dwarf woody species with clonal spread (c). This collection is from cerrado in Brazil (a–c), and grasslands in South Africa (d–f). a. *Psidium grandifolium*, Brazil, b: *Pfaffia jubata*, c: *Bauhinia dumosa*, d: USOs of *Gnidia*, above, *Clutia*, below; e: *Raphionacme*; f: *Eriosema* emerging from a sod of grassland. Courtesy of a,b, Giselda Durigan, c, Alessandra Fidelis, d–f, Nick Zaloumis. Refer to Fig. 4.11 on page 58.

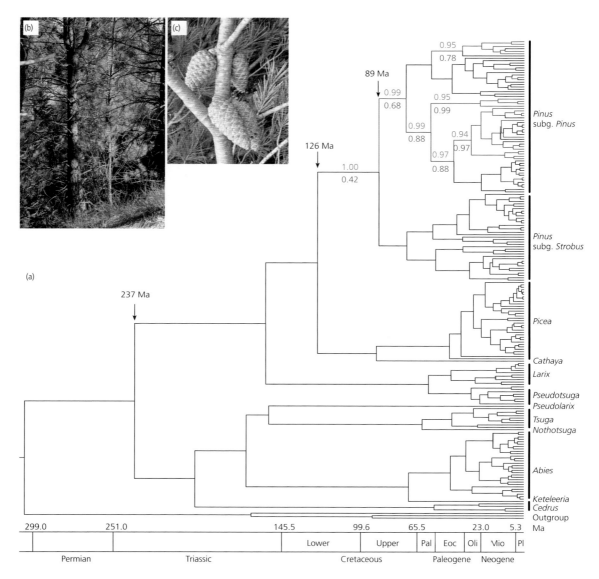

Plate 10 Phylogenetic evidence for novel fire regimes in the Cretaceous selecting for serotiny in *Pinus*. The dated molecular phylogeny indicates the evolution of closed cones, opening after crown fires, at ~89 Ma, in the mid-Cretaceous. (From He et al. (2012), *New Phytologist*.) Refer to Fig. 5.5 on page 72.

(a)

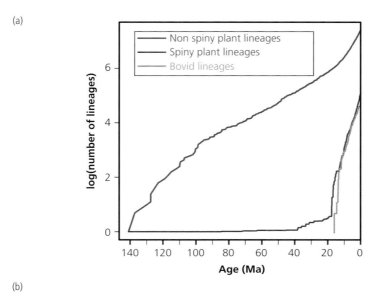

(b)

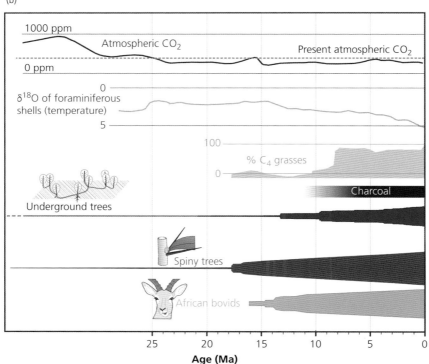

Plate 11 Factors influencing development of savannas in Africa. (a) Diversification of bovids (including antelope) and spiny plants. Proboscideans, including elephants, and hyracoids, dominated the herbivore assemblages in the Cenozoic until bovids entered Africa from Eurasia ~16Ma. (b) Spiny plants prefer open habitat, so that bovids may have opened up forests for savannas before fires triggered C_4 grass expansion in the Late Miocene. From Charles-Dominique et al. (2016). Refer to Fig. 5.10 on page 79.

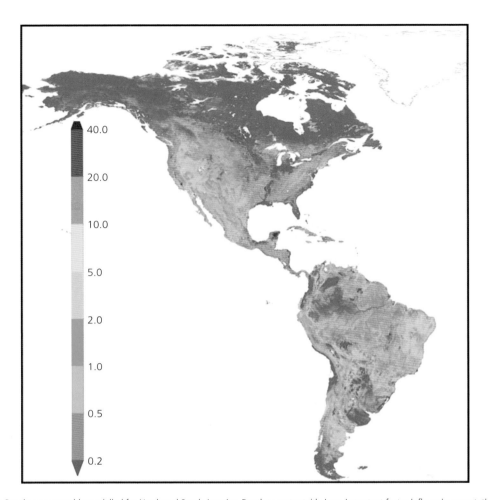

Plate 12 Depth to water table modelled for North and South America. Depth to water table is an important factor influencing vegetation structure, hydrology, drought response, and weathering rates. The map shows estimated mean maximum depth of root water uptake derived from inverse modelling. It suggests striking differences in rooting depth in temperate and boreal regions compared to the tropics. Within the tropics, the maps point to regions of possible edaphic control (blue, seasonally flooded sites favouring open grassy systems versus better drained soils more suitable for forests). Map from Fan et al. (2017). Refer to Fig. 6.3 on page 86.

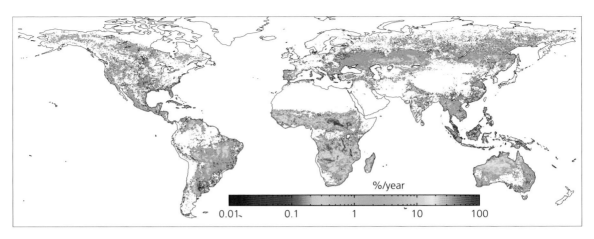

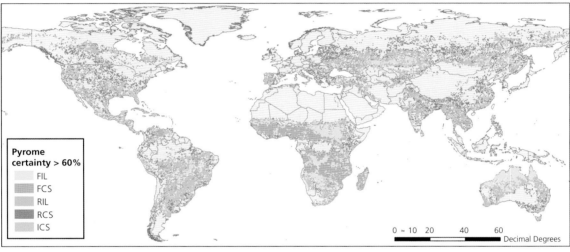

Plate 13 Top: Global burnt area, mean values from 1997 to 2009. From Giglio et al. (2013). Refer to Fig. 7.1 on page 101.
Bottom: World pyrome map with units based on shared fire frequency, intensity and size. From Archibald et al. (2013). Refer to Fig. 1 in Box 7.1 on page 99.

Plate 14 In this mature jack pine stand (*Pinus banksiana*) the surface fire on the left is transitioning to a crown fire on the right. Jack pine is killed by crown fires but recovers from seeds stored in serotinous cones. Fire near Fort Providence, Canadian North-West Territories. With thanks to Stefan Doerr. Refer to Fig. 7.3 on page 103.

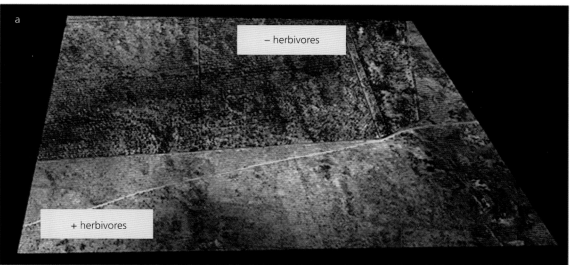

Plate 15 Top: Effect of elk browsing in Yellowstone National Park. a) An aspen stand showing heavy bark damage from elk along the lower several metres of each tree and a long-term lack of recruitment with only large diameter trees present. b) High recruitment within a fenced exclosure. Root sprouts are present outside the exclosure but browsed short by elk. From Beschta & Ripple (2009). Refer to Fig. 8.1 on page 122. Bottom: Airborne 3-D imaging of herbivore treatments in the Kruger National Park, South Africa. The treatments have been applied since the 1970s with a mesh fence enclosing low numbers of rare antelope but excluding all other herbivores larger than a hare. Tree breakage outside the exclosure indicates that elephant browsing is the main agent transforming the vegetation. Long-term basalt site (Nwashitsumbe). Colour-infrared spectroscopy highlights vegetation canopies (red) and dry/senescent vegetation and bare soil (blue to grey) overlain on the 3-D structure of each woody plant at a spatial resolution of 56 cm. From Asner et al. (2009). Refer to Fig. 8.6 on page 127.

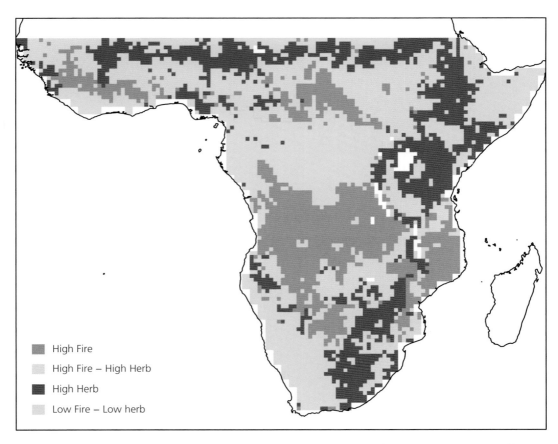

Plate 16 Consumer realms for Africa. High fire = black world, high herbivory = brown world, low fire—low herbivory = green world. Mixtures of black and brown world at the mapping scale are also shown. Note that 'green world', at equilibrium with climate, includes tropical forest but also arid shrublands in the south-west and north-east that are too dry to support closed woody ecosystems. From Archibald and Hempson (2016). Refer to Fig. 8.13 on page 137.

for vegetation structure. This is the case for the Florida coastal plain discussed by Noss (2012). Similarly, erosion gullies in Namibian rangelands have lowered the water table allowing incursion of woody plants (Pringle et al. 2011).

Seasonally waterlogged soils are typically characterized by pale sandy surface soils overlying a mottled, generally more clay-rich subsoil. The surface soils, stripped of clay by lateral water movement (leaching), are often poor in nutrients. Such soils are characteristic of large areas of the Espinhaca range of central Brazil where they are derived from hard, erosion-resistant quartzitic soils. They support grasslands or sparse savannas ('campos rupestris'; Fernandes 2016). Closed forest patches do occur in these grassland landscapes on deep red (freely drained) soils. Here fire suppression is unlikely to lead to the forest spreading into the grassland because of seasonal waterlogging, a strong edaphic control on tree growth (Figure 6.1). In a rare study focussing on depth to water table as a constraint on trees, Leite et al. (2018) studied woody thickening in a mosaic of grassland, savanna, and closed woodland in Brazil. Though the area had been protected from fire for more than 20 years, seasonally waterlogged areas showed little change in woody populations. Leite et al. drilled a network of wells and were able to link depth of the water table in the landscape to the density and diversity of trees (Figure 6.2). Since the soils all had rather similar soil chemistry, they argue that depth to water table was the key factor controlling vegetation structure in their study region. Their results contrast with Durigan and Ratter (2016) who reported widespread woody thickening where fires had been suppressed. It would be very interesting to explore long-term trends under fire suppression on different soil types. Woody thickening should be most pronounced on freely drained soils (red or yellow subsoils) and least likely on seasonally waterlogged soils (grey and mottled subsoils).

6.2.2 Maximum rooting depth as a global constraint on trees

Fan et al. (2017) have placed soil drainage patterns in a global context. They note that plant rooting depth influences resilience to drought, connects deep soil/groundwater to the atmosphere, influencing the hydrological cycle, and controls plant weathering of bedrock, which is important for the long-term carbon cycle. Plant rooting depth can vary greatly within a landscape depending on topography. Plants on deep soils of uplands exploit a root zone fed by depth of infiltration of local precipitation. On mid-slopes, and lowlands, rooting depth is constrained by the depth to the water table with strong inhibition of root growth by anoxic conditions in waterlogged sites. From their large data set, the best predictor of rooting depth was depth to the water table followed by depth to soil barriers such as hard rock or duripans. Fan et al. (2017) used inverse modelling to determine the depth of root-water uptake. They calculated the soil water supply

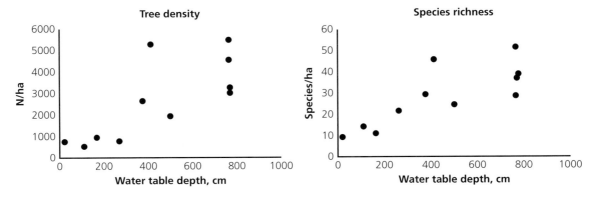

Figure 6.2 Effect of depth to groundwater on tree density and tree species richness in a savanna in Itirapina, south-eastern Brazil. Minimum depth to water table is shown based on fortnightly measurement over two years. Data from Leite et al. (2018).

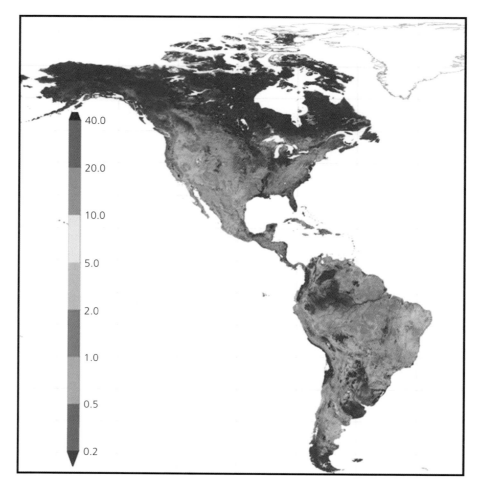

Figure 6.3 Depth to water table modelled for North and South America. Depth to water table is an important factor influencing vegetation structure, hydrology, drought response, and weathering rates. The map shows estimated mean maximum depth of root water uptake derived from inverse modelling. It suggests striking differences in rooting depth in temperate and boreal regions compared to the tropics. Within the tropics, the maps point to regions of possible edaphic control (blue, seasonally flooded sites favouring open grassy systems versus better drained soils more suitable for forests). Map from Fan et al. (2017). (See Plate 12.)

profile from climate, soil texture, and topography using a hydrology model that tracks the landscape distribution of soil water, groundwater, and wetlands. They used satellite-derived leaf area to estimate water demand for the vegetation. Finally, they estimated the rooting depth needed to meet that demand. The result is the first global map of expected plant rooting depth. North temperate and boreal regions have strikingly shallow and homogenous rooting depth estimated in this way. The tropics, in contrast, have complex mosaics of shallow and deep rooting depths depending on topographic position

(Figure 6.3; Plate 12). The production of the map does suggest that the utopian ideal of a global map of soil depth as a constraint on forest development is within reach!

6.3 Chemical controls on forest development

6.3.1 Correlative studies

The absence of forests on many nutrient-poor soils has been noted in different geographic regions and under widely differing climates. In Australia, Beadle (1954, 1966) noted a correlation between soil phosphorus, leaf scleromorphy, and vegetation stature with closed forests restricted to P-rich soils. Since his seminal contributions, Australian vegetation patterns have often been interpreted as being determined by soil nutrient status with sclerophyllous, 'open' vegetation on low-nutrient soils and closed thicket or forest vegetation on nutrient-rich soils (Specht 1979; Westoby 1988; Bowman 2000). Some of the most P-deficient soils in the world occur in south-western Australia and the south-western Cape regions of South Africa (Specht & Moll 1983). Closed forests are rare in the Cape region (Chapter 1) and do not exist in south-western Australia despite areas of high rainfall. Both regions support heathland vegetation dominated by sclerophyll shrubs but trees (eucalypts) are prominent in parts of south-western Australia. The generally low vegetation stature has been attributed to low soil P (Cowling & Campbell 1980; Specht & Moll 1983). It has been argued that the remarkable plant species richness and endemism of these areas is also linked to their low soil nutrient status (Hopper & Gioia 2004).

Hopper (2009) has generalized this argument and proposed a syndrome of plant ecological traits in 'Old, Climatically Buffered Infertile Landscapes' termed OCBILs for short. These are contrasted with YODFELs, 'Young, Often Disturbed, Fertile Landscapes' mainly in the Northern Hemisphere. Besides the south-western Australian and Cape fynbos shrublands, other candidates are the Pantepui highlands of Venezuela and, some argue, the campos rupestre of Brazil (Silveira et al. 2016). All of these 'OCBILs' experience frequent fires unlike the YODFELs of the northern hemisphere. As noted in Chapter 1, forests in fynbos are very different from the flammable fynbos shrublands so presumably a long history of fire has to be added to the mix of OCBIL features (Mucina & Wardell-Johnson 2011).

A correlation between nutrient-poor soils and low-statured vegetation has also been noted for tropical and subtropical savannas and grasslands, especially in South America (Lloyd et al. 2008). In Brazil, tree height and tree basal area vary with soil nutrients, especially P, increasing from grasslands, through cerrados (= savannas), with highest tree biomass and highest soil P in forested sites (e.g.

Goodland & Pollard 1973). Many cerrado plants have low tissue nutrient concentrations and high aluminium concentrations, believed to be characteristic of their low nutrient soils (see e.g. Haridasan 1992; Ruggiero et al. 2002). Thus, as in the heathlands of Africa and Australia, it has been argued that cerrado soils are too nutrient poor to support forests (Cole 1986; Goodland & Pollard 1973; Haridasan 1992). Grasslands in high rainfall regions elsewhere in South America have also been attributed to nutrient-poor soils (Fölster et al. 2001; Huber 2006; Lloyd et al. 2008; Sarmiento 1992). Similar arguments have been applied to high rainfall African savannas, many of which also occur on ancient, deeply weathered land surfaces with nutrient-poor soils (Cole 1986).

If low soil nutrient availability accounts for the absence of forests and their replacement by savannas, then you would expect strong correlations between biomass and soil nutrients within tropical forests. However, in an early review of forest nutrition, Vitousek and Sanford (1986) concluded that 'an association between soil fertility and above-ground biomass is…unlikely in any but the most extreme cases'. Anderson and Spencer (1991) noted that 'there appears to be little evidence that the stature and productivity of mature forests are related to the inherent fertility of parent soil'. Studies both for and against positive correlations of above-ground biomass and soil nutrients have been reported (e.g. Kitayama & Aiba 2002; Quesada et al. 2009, Takyu et al. 2003; Laurance et al. 1999; Paoli et al. 2008). It has been suggested that rates of tree mortality and recruitment (demographic turnover), rather than plant growth rate and resource supply, limit forest biomass (Midgley & Niklas 2004; Keeling & Phillips 2007). If this were true, then simple linear relationships between biomass and soil nutrients would be unlikely.

6.3.2 Plant–soil feedbacks

A major problem with studies of soil/vegetation patterns is distinguishing between the intrinsic properties of the soil and those derived from the vegetation growing on it. Are forests restricted to nutrient-rich soils or are the soils nutrient-rich because of the forests? How important are plant–soil feedbacks in changing the properties of soils in forest/open ecosystem mosaics?

There are many studies on plant–soil feedbacks, especially in northern temperate and boreal regions (reviewed by Ehrenfeld et al. 2005; van der Putten et al. 2013; and see Kulmatiski 2018 for a 7-year field experimental study). An excellent example of how plants can modify soil is the effect of mosses in the genus *Sphagnum* (van Breemen 1995). A combination of polyuronic acids produced by the mosses and slow decomposition of plant tissues acidifies soils to pH values <4.5. Within a few years of establishment, *Sphagnum* causes soil pH to drop by two or more units. The acidity contributes to elimination of plant competitors unable to tolerate the acidic conditions. *Sphagnum* not only acidifies soils, but the morphology of the mosses produces raised bogs, saturating the soils and altering the structure of the soil surface. The low bulk density inhibits root growth of vascular plants. Peat accumulates, providing a major barrier for plant roots to penetrate through to the mineral soil. Consequently, *Sphagnum* can transform mineral soils which support tree growth to peaty bogs that prevent forest growth.

There are many examples of trees modifying soil properties and such modification is considered a key feedback helping to maintain forest or open ecosystems (Wilson & Agnew 1992). The podsolization process is a well-known example. Conifers and heathland plants promote podsolization while broadleaved deciduous trees inhibit it (Nihlgård 1971). Indeed, some long-lived tree species, such as the conifer *Agathis australis* in New Zealand, can produce podzols underneath the canopy during the life of the tree (Wyse et al. 2014). Foresters in temperate regions are well aware of changes to soil properties due to planting of different tree species on the same soil type. Multi-decadal experiments have been conducted in which different tree species are planted on the same soil type to detect species effects on changes in soil chemistry (Ovington 1953; Binkley & Valentine 1991; Binkley & Giardina 1998; Hagen-Thorn et al. 2004). The results of a multi-site experiment with similar treatments are shown in Figure 6.4.

In tropical systems, soils also change markedly beneath a tree when it recruits into grasslands (Ludwig et al. 2004; Coetsee et al. 2010). Indeed, the effect of a tree on soil properties (organic matter content, nitrogen) far exceeded the effect of 50 years of savanna burning in Kruger National Park (Coetsee et al. 2010; Holdo et al. 2012). Hydraulic lift from deeper soil layers to the topsoil lifts nutrients to the surface as well as soil moisture. Time series of soil nutrient changes in fire exclusion plots would be particularly interesting to analyse, especially where fire suppression leads to forests replacing savannas.

Given that plants can modify soil nutrients, depending on their litter quality and the microclimate they create, nutrient differences between forests and open systems may often be an *effect* and not a *cause* of the biome shift. Just how large the effect is, and for what nutrients, has been explored for global soil data bases by Jobbágy and Jackson (2001) (Figure 6.5). Those nutrients most in demand by the plant accumulate in surface soil layers. Therefore, the properties of surface soil layers are those most

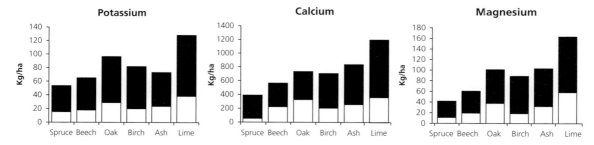

Figure 6.4 Effects of temperate trees on soil nutrients grown on the same soil for 30–40 years. Trees were grown on former arable land at five different sites (Lithuania, Sweden, Denmark). Lower layers (white; 20–30 cm) did not differ significantly in nutrient content. Upper layers (black, 0–10 cm) showed significant differences with greatest contrast between spruce (*Picea abies*) and lime (*Tilia cordata*). Data from Hagen-Thorn et al. (2004).

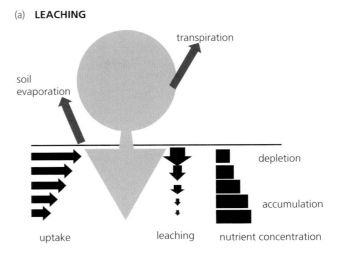

(a) **LEACHING**

transpiration

soil evaporation

depletion

accumulation

uptake   leaching   nutrient concentration

(b) **PLANT CYCLING**

above-ground recycling (litter + throughfall)

transport

uptake

accumulation

depletion

below-ground recycling

nutrient concentration

Figure 6.5 Effect of vegetation on soil chemical properties: (a) shows the effects of leaching on the distribution of a nutrient down a soil profile; (b) shows the effects of plant uptake of a nutrient with re-distribution to surface layers by litter and through fall. Nutrients most in demand should accumulate in the surface soil layers relative to deeper layers, countering the effects of leaching. The relative concentration of a nutrient in the surface soil layers is a useful indicator of differences in nutrient demand in different ecosystems. Figure redrawn from Jobbágy & Jackson (2001).

strongly influenced by the vegetation growing at a site. Nutrients least in demand would be similar down the profile or, if readily leached, may have increasing concentrations down the profile. Jobbágy and Jackson (2004) tested these effects in *Pinus pinaster* plantations planted on coastal dune sands over a 55-year period. Potassium showed increasing concentrations in the surface layers after 15 years and strong increases to a depth of 50 cm after 55 years. Sodium (not an essential nutrient), in contrast, showed decreases in the surface layers consistent with leaching due to rainfall infiltration (Figure 6.6). This graphic demonstration of plant effects on

nutrients, and vertical nutrient distribution in the soil, shows how powerfully plants can alter soil nutrient supply.

These effects extend, too, to micronutrients. Jobbágy and Jackson (2004) also report a chronosequence of changes in Mn concentrations when Pampas grasslands are converted to eucalypt plantations. Within 50 years, eucalypts had dramatically altered the distribution of Mn, increasing concentrations by more than tenfold in surface soils while halving concentrations at depths of 20–60 cm (Figure 6.7). Iron concentrations, in contrast, varied little. Both K and Mn have been implicated as nutrients controlling forest

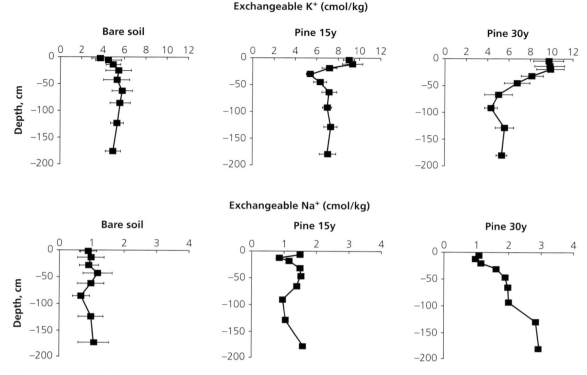

Figure 6.6 Vertical distribution of exchangeable base cations along a chronosequence (0–30 yr shown) of *Pinus pinaster* plantations in dunes of coastal Argentina. The bare dunes were stabilized with dry tree branches five years before planting, with measurements obtained at the end of this stabilization period. Potassium (top) was concentrated in the surface layers by the pine with the concentration layer deepening over time. Sodium (bottom) was depleted from surface layers by leaching. Redrawn from data in Jobbágy & Jackson (2004).

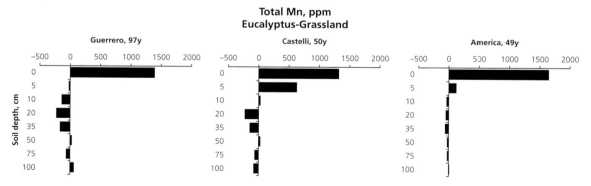

Figure 6.7 Vertical distribution of manganese expressed as the difference in mean values under *Eucalpytus camaldulensis* versus native pampas grassland at three sites in Argentina. The plantations are 97, 50, and 49 years old, grow in a humid to subhumid temperate climate, and have mollisol soils that were never ploughed, fertilized, or irrigated. From data in Jobbágy & Jackson (2004).

development (Mills et al. 2016, 2017; Lloyd et al. 2015). But these dynamic changes in K and Mn suggest it is the plants, not the soils, that are determining soil nutrient supply by redistributing mineral elements from deeper to shallower soil layers.

6.4 Sampling soils for ecological studies

How deep should one sample soils to estimate potential nutrient pools for supporting forests? The common ecological practice of sampling only the surface soil layers to characterize nutrients at a site confounds the intrinsic chemical properties of the parent material with the biotic forcing of soil properties by the plants that grow there. While global surveys indicate most root biomass is in the surface layer of soils (Jackson et al. 1996), roots explore much deeper soil layers. From a global survey in diverse vegetation types, Canadell et al. (1996) reported mean maximum rooting depths of 7.0 m for trees, 5.1 m for shrubs, and 2.6 m for herbaceous plants. Though most nutrient uptake is probably from surface soil layers, transpiration-driven mass flow is a mechanism for accessing nutrients in deeper soil layers (Cramer et al. 2009).

Thus, the most likely place to find nutrient-limited open vegetation in a forested landscape would be on shallow, leached soils with no possibility of root growth to access deeper soil volumes. Such soils are often seasonally waterlogged, or drought-prone, and it may be difficult to disentangle the role of nutrient limitation from impeded root development in excluding forests (see e.g. Lloyd et al. 2008; Leite et al. 2018).

6.4.1 Which nutrients?

Nitrogen is intensively studied in biogeochemical cycling, but its source is the atmosphere and not geological substrate. Phosphorus has also been intensively studied and has long been considered an important element determining major vegetation patterns. However low P availability has often been inferred from sclerophylly but not measured in the soil, particularly in Australia. A recent survey of soil P by Kooyman et al. (2017) has shown that low P availability has been exaggerated for most of

Australia. At least a third of the continent has soil P in the range 101–250 mg kg^{-1} which is unexceptional. Lehmann et al. (2011), in comparing savannas across the southern continents, reported that Australian soils were not as nutrient-poor as many African or South American savannas. And Cramer et al. (2018) have shown that extremely P deficient soils derived from quartzites in the south-western Cape of South Africa support sclerophyllous fynbos vegetation but also closed forests. The forests occur in the same landscapes, on the same soils, with the same sand texture, but have at least five times more total P than adjacent fynbos soils. Fynbos, it seems, creates its own miserably low P soils.

Calcium has been the focus of several studies on sustainability of tropical forest harvesting since it frequently appears to be a limiting nutrient (Nykvist 2000). Ca stocks are so limited in some tropical forests that it is difficult to understand how these forests exist. Nykvist (2000), for example, has argued that trees must be accessing not only exchangeable Ca but a large fraction of the total Ca (which is ~double the exchangeable in his analyses) to explain biomass in the forest. Several studies have shown that trees 'mine' the rock directly and can break down rock minerals to extract Ca (Bormann et al. 1998). K also appeared to be limiting forest biomass in at least as many instances as Ca. The disparity between exchangeable and total K is even larger than Ca with five times more total than exchangeable amounts in some studies (e.g. Wilcke & Lilienfein 2004). This, together with atmospheric inputs, sometimes from very distant sources (e.g. Boy & Wilcke 2008 for Andean forests; Soderberg & Compton 2007 for fynbos; Vitousek 2004) might help explain how forests persist despite the apparent K deficiency in many soils. Disappointingly few studies reported the necessary information to analyse nutrient stocks in a way one can estimate potential woody biomass. To do so, you need to sample to at least 1m depth to help determine the extent to which plants are concentrating nutrients (Figure 6.5; Jobbágy & Jackson 2001, 2004). Secondly, you need to determine volumetric measures of nutrients, and not merely nutrient concentration, to estimate availability on an area basis (Wigley et al. 2013).

6.5 Why should nutrients influence forest/open ecosystem distribution?

Debates on the relationship between soil nutrients and forest biomass have been almost entirely based on correlations. But correlations cannot reveal potential vegetation; they cannot show whether a site has the potential to support forests that could replace 'early successional' grasslands, savannas, and shrublands. Potential vegetation at a site should, in principle, be predictable from ecosystem physiology. DGVMs were developed to test physiological knowledge and the extent to which physiology explained major vegetation patterns (e.g. Haxeltine & Prentice 1996; Woodward & Lomas 2004). The current generation of global vegetation models does not incorporate soil nutrients (other than nitrogen which is seen as a biologically controlled nutrient accessed from the atmosphere; Cramer et al. 2001). Yet correlations of vegetation patterns with soil nutrients suggest that soil chemistry is an important factor influencing vegetation properties (Beadle 1966; Goodland & Pollard 1973; Lloyd et al. 2008; Specht 1979). The challenge is to develop a sufficiently coherent mechanistic explanation for nutrient dependence of different growth forms to include in a global simulation model.

6.5.1 Hypotheses for soil nutrient dependence

Rather surprisingly, there have been few explicit hypotheses as to why soil nutrients should limit development of forest in the first place. One possible explanation is that forests have a higher demand for nutrient stocks than low biomass systems because of nutrients locked up in wood. Bond (2010) quantified the minimum soil nutrient stocks required to build a representative tropical forest with aboveground biomass of 350 Mg ha^{-1}. Using values typical of Amazon forests for nutrient content of leaves and wood, he estimated that tropical forests would need a minimum stock (kg ha^{-1}) of 20–30 P, 200–350 K, 300–600 Ca and 55–65 Mg. This minimum requirement would not be met from the surface layers of cerrado soils for calcium and potassium. However, if plants accessed deeper soil layers (to 1m or more) they would obtain sufficient nutrients, including calcium and potassium, to build a forest. Since the mean maximum rooting depth of trees, globally, exceeds 7 m, Bond concluded that nutrient stocks would very seldom limit forest development except where soils are very shallow. But shallow soils also restrict available root space and soil moisture for tree growth. Thus, the nutrient stock hypothesis fails to explain correlations between the distribution of forests and nutrient-rich soils.

6.5.2 Nutrient constraints on tree growth

Mills, Milewski, and colleagues (Mills et al. 2016, 2017) have proposed a nutrient-based metabolic mechanism for why grasslands prevail where the climate can support forest. Their hypothesis is that plant metabolism can be altered by micronutrient availability to favour catabolism (breaking down complex molecules, e.g. respiration) or anabolism (building up molecules, e.g. photosynthesis, protein synthesis). Trees require more materials than grasses to build their stems from carbon-rich lignin, cellulose etc., and so should require proportionately more minerals supporting anabolic processes. Where soil nutrients supporting catabolic processes exceeded those supporting anabolism, then metabolism would not produce enough material to build trees, and grasses would dominate. Most nutrients are involved in both catabolic and anabolic processes but some, Mills et al. suggest, are used either for catabolic or anabolic metabolism. As a preliminary categorization, Cu, Zn, and Se were classed as mainly catabolic and Mg, Fe, and Mn as mainly anabolic. Thus, trees should require more Mn than grasses to build structures from anabolic processes. Interestingly, this prediction is supported for eucalypt woodlands versus grasslands in Argentina where the trees rapidly began accumulating very large quantities of Mn compared to grasses (Figure 6.7). However, Mills et al. (2016) suggest that ratios of catabolic/anabolic soil nutrients might be better predictors of grassland versus forest supporting soils. They predicted that the ratio of Cu/Mn should predict whether a soil supports grasslands (a high value) or woodlands (a low value). In a comparison of soils in Australian woodlands versus grasslands they found mean values of Cu/Mn concentrations of 0.03 in grasslands versus 0.019 in woodlands and considered these differences sufficient to support their hypothesis.

Evidence for the catabolism/anabolism hypothesis is correlative and based on sampling soils with and without trees. Experiments manipulating Cu/Mn and observing how different ratios might influence grass versus tree growth have yet to be done. Mills et al. (2016) only sampled the top 10cm of soil, the layer most influenced by the plants themselves (see Figure 6.7 for Mn). Hagen-Thorn et al. (2004), in their study of feedbacks of different temperate tree species to soil mineral properties, sampled the surface 10 cm and the subsoil (20–30 cm) layer beneath that. Topsoils were enriched with Cu relative to subsoils, ranging from 1.2 to 2.2 times more for oak and spruce respectively, and for Mn 1.8 to 4.7 times more in the topsoil versus subsoil for lime and birch respectively. The ratio of Cu/Mn ranged from 0.019 to 0.052 for subsoils (spruce, beech) and 0.013 to 0.027 (birch, beech) for topsoils. This study of long-term effects of trees on soils suggests that the catabolism/anabolism hypothesis suffers the same problem as all the other correlative studies of nutrient effects: are the soil mineral concentrations the causes or consequences of the vegetation growing on them? The many studies on soil–plant feedbacks suggest they are the consequences of vegetation feedbacks, especially when sampled in the surface layers.

Lloyd et al. (2015) have argued vigorously that soil nutrients determine forest/savanna mosaics in Brazil. In their case, potassium was the key variable correlating with tree cover. They have also attempted to devise a process-based argument for why K should be important for trees in this system. However, the study suffers the same problem of correlative evidence and no experimental studies have been developed to test the direction of causality. The Jobbágy and Jackson analysis of changes in K concentrations in dune sands planted to conifers suggests that it is the plants calling the shots.

6.5.3 Nutrient/fire interactions

Nevertheless, boundaries between closed forests and open ecosystems often coincide with geological boundaries. In humid climates, forests are typical of clay-rich soils and open ecosystems of coarse textured (sandy) soils which are poorer in nutrients. Kellman (1984) suggested that the explanation for this common pattern was higher tree growth rates on the nutrient-rich soils which allowed trees to recover more rapidly after fire than on nutrient-poor soils. It seems very reasonable that the interaction between soil factors influencing primary production and rates of disturbance should limit forest distribution. Higgins et al. (2000) for savannas, and Hoffmann et al. (2012a) for forests, extended this idea developing a conceptual model that builds on plant growth rates to reach key demographic thresholds. For savanna trees, a key threshold is to reach fire-tolerant sizes, while for forest trees reduced damage is also important, but the critical size threshold is to suppress fire altogether. These fire resistance and fire suppression thresholds (Hoffmann et al. 2012a) are discussed further in Chapter 7.

6.6 Summary of soil effects

Much of the polemic over whether soils or consumers account for vegetation pattern is centred in the tropics and in the southern hemisphere. Similar heated arguments took place between proponents of bottom-up resource-based hypotheses versus top-down trophic control of productivity in freshwater lakes (Carpenter & Kitchell 1988). Apparently, it is now widely accepted that bottom-up explanations fit best to very nutrient-poor and very nutrient-rich lake systems while top-down processes are common in between. The advantage of aquatic systems is that experiments yield very rapid results. In terrestrial environments, experiments take longer. Manipulating relevant consumers, fire and large herbivores, or planting trees to measure responses to soil properties takes decades to give results. As the issues become clearer, we should be better able to design the critical experiments or use existing ones more profitably. The contribution of large-scale mapping, remote sensing, and modelling is also beginning to show the potential for resolving soil effects at a global scale (Fan et al. 2017). Below-ground processes are intrinsically difficult to study, and new technologies may ease the pain of digging to make the observations.

BOX 6.1 Competition: can understorey herbs and shrubs prevent forest development?

The grasses, herbs, or shrubs that dominate the 'understorey' of OEs are seldom seen as the agents responsible for excluding forest. As noted in Chapter 2, asymmetric competition for light means that taller plants shade out smaller ones, so you would expect trees to dominate in late successional communities. But grasses and shrubs may outgrow tree roots in exploring the soil. Thus OE 'understorey plants' could suppress tree recruitment by below-ground competition (Sankaran et al. 2004; Bond 2008). For many years, the plant ecology of savannas was dominated by the question of how trees and grasses coexist. Walter (1971) hypothesized that grasses exploit the upper layers of the soil and tree roots the lower layers. Differential use by grasses of surface layers, and trees of deeper layers, would allow coexistence by niche differentiation. Numerous subsequent studies have shown that grasses compete intensely with trees, especially in their juvenile life stages when seedlings and saplings exploit the same soil layers (Sankaran et al. 2004; Riginos 2009; Scholes & Archer 1997). In experimental studies of resource constraints on plant growth, grass removal treatments usually have by far the largest effects (Figure 1; February et al. 2013). Reduced growth rates mean that juvenile trees will be exposed for much longer to stem losses to consumers, and climate extremes such as frost and drought. Competition with the understorey plants therefore adds greatly to the probability that trees will fail to grow to threshold sizes where they are freed of losses to consumers and buffered against short-term climate catastrophes. This combination of suppression of recruitment rates, and prolonged exposure to risks of remaining small, is widespread wherever growing conditions, especially light, allow vigorous understorey growth. In temperate and boreal forests, for example, heavy deer browsing has helped open canopies allowing vigorous growth of graminoids and ferns. The combination of browsing and herbaceous composition can switch the system from forest to a wooded grassland (Côté et al. 2004; Bradshaw & Waller 2016; Tanentzap et al. 2011). In savannas, the first

response to suppression of fire, or protection from herbivores, is the release of existing suppressed saplings. Recruitment of new individuals can be delayed for years because of poor recruitment (Midgley et al. 2010; Clarke et al. 2013).

Given the ubiquitous suppression of woody recruits by herbaceous and woody low-growing plants, are there circumstances where understorey competition excludes trees? Jurena and Archer (2003) provided a useful conceptual framework for addressing the question. Using the analogy of canopy gaps as prerequisites for tree recruitment in forests, they suggested that tree seedlings could only recruit successfully in 'root gaps' in grassy swards. By creating root gaps, they were able to show that deeper gaps were more important than wider ones, that above-ground grass biomass was poorly correlated with below-ground root density, and that root gaps did indeed contribute greatly to recruitment success (Figure 2). Vandenberghe et al. (2006), using similar arguments, tested the importance of gaps and gap sizes in the recruitment of tree seedlings in alpine meadows. They too found that gaps greatly enhanced recruitment probabilities but that larger gaps led to desiccation of the seedlings in the particularly hot conditions during their experiments.

Wakeling et al. (2015) explored the root gap hypothesis as an explanation for vegetation change along an elevation gradient in South Africa. Savannas occur at lower elevations, switching to treeless grasslands at high elevations. Interestingly, although the grasslands are treeless, they co-occur with closed forests in the same landscape. They tested the prediction that root gaps should be common in savannas, rare as tree densities decline, and absent in the upland grasslands. The frequency of root gaps, measured by sampling roots in grids of soil samples along an elevation gradient, was highest in the savannas, declining with elevation. There were no root gaps in the grasslands (Figure 3). Parallel surveys of tree seedlings and juveniles showed a similar decline with no seedlings at all in the grasslands, despite the presence of

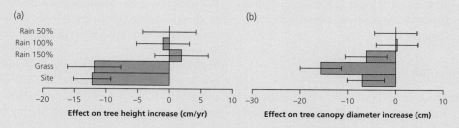

Figure 1 Grass removal treatments had the largest impact on juvenile tree growth relative to rainfall addition and removal experiments at two contrasting sites in an African savanna. From February et al. (2013).

Box 6.1 *Continued*

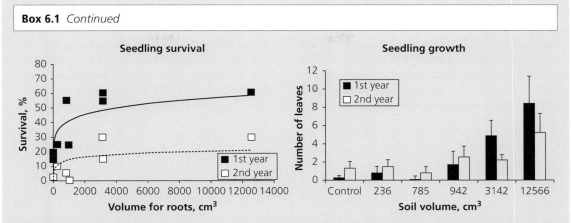

Figure 2 The effect of below-ground gap size on mesquite (*Prosopis glandulosa*) seedling survival (left) and growth (right). Seedlings were grown in tubes of different soil volumes thrust into the soil in a savanna, central Texas, USA. First year was wet, second year drier. From Jurena & Archer (2003).

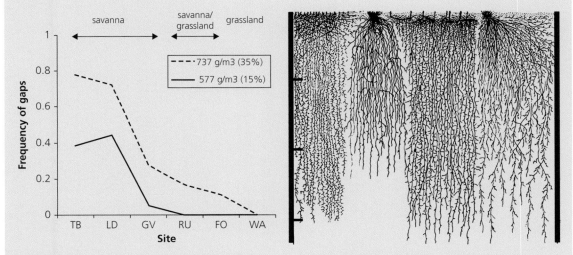

Figure 3 Changes in the frequency of root gaps along an elevation gradient from savannas at lower elevations to montane grasslands at higher elevations, South Africa. Grass roots were sampled in a grid of 18 auger holes at each of six sites. The frequency of gaps at each site is shown for the 15th and 35th percentile of root biomass for all 108 samples. Gaps were frequent in the drier savannas but were absent from the grassland site. From Wakeling et al. (2015).

Figure 4 Profile diagrams of grass root systems in North American mixed prairies. From left to right, *Buchloe dactyloides*, *Aristida purpurea*, *Agropyron smithii*, *Bouteloua curtipendula*. Scale is 0.5 metre intervals to 1.5m. From Weaver (1958).

potential seed sources. Just why grass root gaps disappear in mesic montane grasslands is not known for this system. Clearly the generality of the pattern needs further study. Do the grasslands of the Konza prairie differ from the oak savannas of Minnesota in the density of root gaps in the grass sward? Certainly, prairie grasses have enormous root systems

(Weaver 1958) developed in the Kansas soils (Figure 4). Are Cedar Creek soils too sandy and nutrient poor for comparable development of roots, leaving a high frequency of root gaps for oak seedling recruitment? Jurena and Archer's root gap hypothesis is also important for exploring resource constraints on tree seedling recruitment in savannas. If tree seedlings generally recruit in gaps with sparse grass roots, then experiments on factors limiting tree seedlings should avoid confounding results by growing seedlings with grasses.

What seems to have been under-appreciated in the past is the extent of plant–soil feedbacks. The implicit assumption of deterministic causality is being challenged. This is the chain of reasoning that, for a given climate, and for a given soil, there will be one vegetation type made up of optimally suited growth forms for those site conditions. Instead feedbacks from the biology of the ecosystems to its physical setting, as postulated by ASS theory, appear to be the norm. Life, it seems, has enormous capacity for modifying the physical environment to suit the resident organisms. The implication is that there are no fixed, optimal solutions to the environment and that, at least in some climates, and on some geologies, there are multiple solutions to the problems of existence, generating multiple ecosystem states.

CHAPTER 7

Fire and open ecosystems

Fire alone, without any other activity, produced three major changes in the forest: opening up the forest, creating grasslands, and altering the composition and range of trees. P.26. Michael Williams, Deforesting the Earth. University of Chicago Press, 2006.

7.1 Introduction

Twenty years ago, one could describe fire ecology as a 'ghetto science', of interest to a few specialists in different regions but ignored by mainstream ecology as shown by lack of content in general ecology textbooks. All that has changed. Today fire ecology is a vast topic with an extensive literature and a global reach. Readers have a wide choice from popular books (Scott 2018) to general introductory texts (Scott et al. 2014; *Fire on Earth*), to detailed regional accounts such as Keeley et al. (2012) for Mediterranean climate regions, Cochrane (editor, 2009) on tropical fire ecology, Bradstock et al. (2012) on 'Flammable Australia' and many more. This chapter is, therefore, not an introduction to fire ecology. The intention is to address the question: To what extent can fire account for the existence and extent of open ecosystems?

For much of the twentieth century, and lingering in popular attitudes today, wildfires were seen as destructive agents of human origin. More than half a century ago, Sauer (1950) wrote an insightful review of the importance of fire in structuring global vegetation still relevant today. However, in keeping with the times, he assumed that fires were not 'natural' but anthropogenic artefacts. By implication, open, flammable ecosystems must necessarily be anthropogenic too. So the most common hypothesis for the existence of flammable open ecosystems for most of the twentieth century was that they are the product of deforestation by human-set fires. If fires were suppressed, the system should

return to its 'original' forest condition. The open ecosystems are merely early successional stages on their way to forests.

In this chapter, I explore the role of fire as consumer maintaining open ecosystems. The spatial extent of vegetation fires is vast, as revealed by satellite imagery, and mostly burns in open ecosystems. Fires differ in their effects depending on the fire regime, the pattern of means and variance in fire frequency, intensity, and fire sizes. Understanding the ecology of fire requires an understanding of fire regimes. The role of fire in maintaining open ecosystems has been studied by experiments, intentional or incidental ('natural'?), where fire has been suppressed for decades. The many examples give an idea of rates of change and environmental context of biome switches to closed forests. Experiments attempting to convert closed forests to open ecosystems are far rarer, perhaps because it is generally assumed that forests are extremely vulnerable to burning. Having reviewed evidence for fire effects on forests, I turn to Alternative Stable State (ASS) theory and the evidence for patterns in tree cover consistent with the theory. Feedbacks helping to maintain the alternative states are discussed, including the ecology and evolution of flammability. The stability of ASS, especially in tropical forest/savanna mosaics, has been studied by palaeoecologists who have observed both stability and sudden regime shifts between forest and OE states over time scales of centuries to millennia. Finally, I consider

Open Ecosystems: ecology and evolution beyond the forest edge. William J. Bond, Oxford University Press (2019). © William J. Bond 2019.
DOI: 10.1093/oso/9780198812456.001.0001

the ecology at the edges of forests and OEs since this is the interface from which forests might advance or retreat, ultimately influencing the fate of the ASS.

7.2 Hypotheses on the role of fire in shaping ecosystems

Paleoecological studies began undermining the hypothesis that fires were all anthropogenic 'distur-bances' by finding evidence for charcoal long pre-dating human settlement and of the existence of ancient shade-avoiding open ecosystems. Acceptance of the antiquity of fire, however, does not necessar-ily mean acceptance of the view that fire is a signifi-cant factor shaping ecosystems. Weather conditions are important contributors to how a fire burns, and climate influences the longer-term pattern of fires (the fire regime, Box 7.1) so that there is a strong abi-otic component to fire as a consumer. The abiotic

---

Box 7.1 Fire regimes

Fire is not a unitary phenomenon but a multi-dimensional eco-logical process. Understanding these dimensions helps under-standing the diversity of fires, and the diversity of vegetation responses. Malcolm Gill first used the concept of fire regime to interpret the diversity of Australia's flammable vegetation (Gill 1975). Since then, the fire regime concept has been devel-oped, modified, and refined (Bond & Keeley 2005; Keeley et al. 2012; Archibald et al. 2013). Fire regimes are characterized by their frequency (the occurrence of fire for an area and period of interest), intensity (energy released and related features such as flame length), severity (the ecosystem impact measured, for example, by biomass consumed), season (especially the length

of the fire season, when vegetation is dry enough to burn), and fire size. The mean, range, and variance in these regime vari-ables are informative. One of the easiest but most informative components of a fire regime to record is the fire type. The type varies from *ground* fires consuming organic soils, *surface* fires which burn just above the ground surface (fuelled, e.g., by grasses, ferns, tree leaf litter), and *crown* fires which burn the canopy of shrubs, or trees if present. Some systems, especially conifer woodlands, have a mixed fire regime of surface, crown, and unburnt patches. These mixed fire regimes are important for maintaining species, such as ponderosa pine, which could not easily survive in either pure surface or crown fires.

Box Table 7.1. Examples of fire regimes in different vegetation types and characteristic plant traits and life history types. After Scott et al. 2014, with added traits and vegetation types.

Fire type	Surface	Surface	Surface	Crown	Crown	Crown
Fuel type	C_4 grasses	C_4 grasses	Leaf litter, twigs shrubs	Woody biomass	Woody biomass	Woody biomass
Fire intensity	Low-Moderate	Low-Moderate	Low	High	High	High
Fire return interval, y	1–3	5–10	5–10	15–100	30–150	50–500
Climatic conditions	humid, wet/dry	after wet years	after wet years	drought, hot	drought, windy	drought, windy
Fire-stimulated recruitment	No	No	Low	High	Moderate	
Epicormic (crown) sprouting	High	Moderate	Low	Low	Low	Low
Basal sprouting (woody)	High	Moderate	Low	High	Moderate	Moderate
Non-sprouting woody plants	No	No	Low	High	Moderate	Moderate
Self-pruning (removing dead branches acting as fuel ladders)	High	Low	High	No	No	High
Thick bark	High	Variable	High	No	No	Low
Fire severity (biomass consumed)	Low	Low-Moderate	Low	Moderate	No	High
Vegetation type	Mesic savanna	Dry savannas	Dry ponderosa pine woodland W. USA	Fire-dependent Mediterranean shrublands	boreal conifer forest North America	Wet eucalypt forest Australia

Different fire regimes select for different plant traits and act as a strong environmental filter (Box Table 7.1.) Similar fire regimes select for similar traits in different geographic regions. For example, there is convergence to similar fire survival traits (sprouters, non-sprouters), recruitment traits (serotiny, fire-stimulated seed germination), and vegetative traits promoting flammability in the Mediterranean-type shrublands of California, Mediterranean matorral, Cape fynbos, and Australian kwongan (Pausas et al. 2004; Keeley et al. 2012). The suite of traits is completely different in the grass-fuelled surface fire regimes in savannas (e.g. Hoffmann et al. 2003, 2004; Charles-Dominique et al. 2018). A switch from one fire regime to another can cause cascading species losses. For example, a fynbos grass, *Ehrharta calycina*, has invaded the heathlands (kwongan) in the vicinity of Perth, Western Australia. The alien grass has switched the fire regime from a crown fire system to one of frequent savanna-like surface fires, destroying the rich diversity of the endemic heathlands (Fisher et al. 2009).

The first attempt to map global fire regimes was made by Archibald et al. (2013) who used a 14-year record of remote sensing of fire by satellites. They were able to measure five variables from the satellite imagery: fire return interval, maximum fire intensity, the length of the fire season, maximum fire size, and mean annual area burned. These fire metrics were a considerable advance on previous indices of burnt area alone. Using a clustering algorithm, they extracted five global syndromes of fire regimes, for which they coined the name 'pyromes', analogous to biomes. Pyromes are defined by fire frequency (Frequent, Intermediate, or Rare), intensity

(Intense, Cool) and size (Large, Small). Two of their global pyromes are associated with surface fires burning grassy fuels: FIL (Frequent, Intense, Large burns) and FCS (Frequent, Cool, Small). The size of fires varies with human population density, small where densely settled, large where sparsely settled. Two pyromes were associated with diverse crown fires in woody fuels: RIL (Rare, Intense, Large) and RCS (Rare, Cool, Small). The fifth pyrome is ICS (Intermediate, Cool, Small) and is primarily anthropogenic agricultural and deforestation fires. At this level of resolution, and for just the five fire variables, pyromes were not closely matched to biomes. For example, the RIL pyrome (Rare, Intense, Large fires) was common in temperate coniferous forests, boreal forests, Mediterranean shrublands, tall eucalypt woodlands (forests), and arid spinifex grasslands in central Australia.

With longer records, especially of fires in woody fuels, and more refined measures of fire activity, the pyrome concept will, no doubt, be refined globally and regionally. The world pyrome map (Figure 1; Plate 13) should be a stark reminder that very different fires occur in different parts of the world. Extrapolating experience, and policy, across pyromes will be as erroneous as extrapolating biological knowledge from one pyrome to a completely different one. The disastrous fires risking lives and property in California, the Mediterranean, and eastern Australia (RIL pyrome) have very different earth-system feedbacks and require utterly different approaches to management than geographic regions dominated by grass-fuelled fires of the FIL and FCS pyromes.

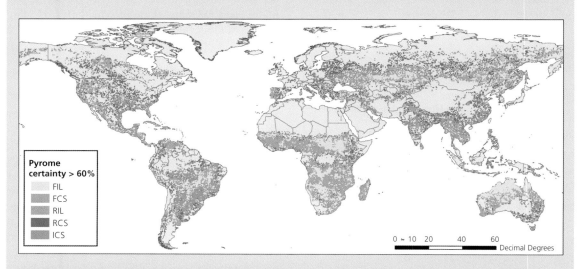

Figure 1. World pyrome map with units based on shared fire frequency, intensity and size. From Archibald et al. (2013). See text for symbols.

hypothesis is, then, that fire is merely a consequence of climate, burning vegetation made of growth forms selected by that climate. The flammability of the ecosystem is merely an incidental consequence of the growth forms selected by the climate and soils. Fire suppression should thus lead to negligible changes in the growth form mix. In contrast, the pyrocentric hypothesis holds that fire is a major consumer shaping the patterns, structure, and composition of vegetation. Flammable ecosystems depend on recurring fires for their existence. Major changes, biome shifts from open to closed ecosystems, would follow, equivalent to trophic cascades, if fires were suppressed or the fire regime altered.

If you read general ecology textbooks, the many pages devoted to competition might suggest that plant competition structures communities. To explain open ecosystems, a competition-based hypothesis would have to argue that low open growth forms (grasses, shrubs) competitively exclude tall trees thereby preventing forests from shading out open vegetation. Though fire may influence the competitive outcome, for example by reducing availability of nutrients, fire suppression would not change the system to a forest.

These hypotheses are not mutually exclusive and for any given landscape, there could be a mixture of them with the relative importance of fire versus site conditions varying over space and time. Fire suppression would then have variable influences on vegetation across a landscape.

Table 7.1 Mean annual burnt area across the world, 1997–2009. From Giglio et al. 2013.

Region	Area Mkm^2	Burnt area Mha	% of world burnt area	% of region that burns
Africa	24.65	242.7	69.5	9.8
Australasia	7.98	50.2	14.4	6.3
Asia	42.77	27.8	8.0	0.6
South America	17.81	21.3	6.1	1.2
North America	19.19	4	1.1	0.2
Central America	2.71	1.8	0.5	0.7
Middle East	12.03	0.8	0.2	0.1
Europe	5.41	0.7	0.2	0.1
Global	132.55	349.3	100	2.6

Not surprisingly, the question of whether fire maintains open systems remains contentious, fed, no doubt, by different local experience and local circumstances. A pyrocentric view of plant communities is that the component parts, the species pool, grouped into growth forms such as trees, shrubs, and grasses, is sorted in different ways depending on fire history. Frequent fires would select for herbaceous plants (with or without thick-barked trees). Less frequent but more intense fires would allow taller shrubs to dominate. Very infrequent fires would allow shade-tolerant closed forests to form. Jackson (1968) first suggested this kind of fire sorting as the process generating different plant communities in Tasmania. Each assemblage contained different species and a different mix of growth forms maintained by positive feedbacks to the fire regime and to the soil. Bowman and Wood (2009) provide a detailed account and critique of the framework. They note that one of the main problems in testing the idea is that the fire return times for the more wooded assemblages are on the order of centuries. These time intervals are too long for direct observation and cannot be studied by dendrochronology (eucalypts do not make annual rings). This problem, of decadal to century-long timescales of vegetation responses to changing fire regimes, is a familiar one to students of vegetation change in terrestrial systems, especially where tree rings cannot be used. Fortunately, there are some fire experiments, intentional or accidental, maintained for many decades which give us some inkling of longer-term dynamics.

7.3 The spatial domain of fire as a consumer

Fires have been mapped by land managers for much of the twentieth century in diverse countries. However, the first global perspective, transcending national boundaries, emerged only in the 1990s with the availability of satellite imagery. Most global analyses in the 2000s are based on MODIS or AVHRR imagery and seldom extend back more than 10 to 20 years. Consequently, vegetation that burns frequently will be well represented but arid and woody systems that burn at multi-decadal or even century timescales will be underrepresented. However the surprise, for many, was the very large spatial extent of global fires (Figure 7.1; Plate 13). Around 3 per cent of the global vegetated land area burns every year (Table 7.1; Giglio et al. 2013). Grassy fuels, and especially savannas, account for most of the burnt area with Africa alone responsible for about 70 per cent of the world's annual burnt area. Australia is the next most flammable region with around 14 per cent of the global burnt area. Europe has the least burnt area (0.2 per cent). Lack of familiarity with fire no doubt accounts, in part, for the general antipathy to fire from European foresters and colonial scientists. Scientific hostility to fire was propagated around the world in European colonies and that hostility is still prominent in global discourse on planetary management (Pyne 1990).

Among landcover classes (= biome types) open ecosystems contribute far more to global burnt area than closed ecosystems (Figure 7.2). Savannas are

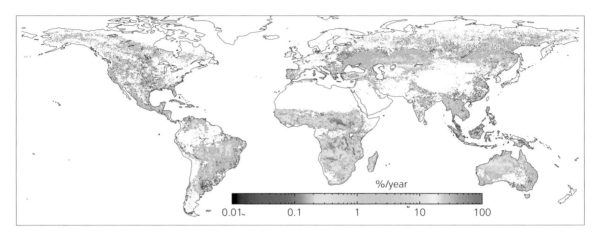

Figure 7.1 Global burnt area, mean values from 1997 to 2009. From Giglio et al. (2013). (See Plate 13.)

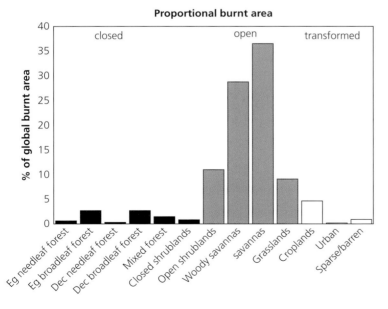

Figure 7.2 Contribution to annual global burnt area from different landcover classes (biome types), 1997 to 2009. Open ecosystems (stippled) burn by far the largest area. From data in van der Werf et al. (2010).

the most frequently burning ecosystem on Earth, including both open ones and more wooded savannas. Savanna woodlands may form nearly closed canopies but continue to burn very frequently if they maintain flammable C_4 grassy understoreys. Confusingly, these savannas have often been called 'forests', especially in south and east Asia. Since forests must be protected from fire according to European forestry custom, these savannas have long suffered from fire protection with unintended toxic consequences (Ratnam et al. 2011). The FAO system of classification, in which savannas with as little as 10 per cent tree cover are described as 'forests', has been particularly damaging in this respect and clearly needs revision to separate out closed forests, as defined in this book, from open ecosystems with shade-intolerant understories (Ratnam et al. 2011; Ratnam et al. 2016; Parr et al. 2014). Besides savannas, an arc of high fire activity occurs across Central Asia from Georgia through Kazakhstan to China. There appears to be no scientific literature (in English, at least) on the fire ecology of the region. Much of the Eurasian fire arc in the west is cropland and in the east steppe/forest transition. So this frequently burnt area also appears to be linked to herbaceous fuels but based on cool temperate C_3 grasses, not tropical C_4 grass clades.

Woody-fuelled fires are less prominent in Figure 7.1. They burn less frequently, and their vital statistics will require longer remote sensing records to quantify patterns. Woody-fuelled fires include boreal forests, Mediterranean climate regions, conifer woodlands in the northern hemisphere, and diverse shrublands, including eucalypt and associated 'sclerophyll' communities unique to Australia. Since these areas also overlap with centres of urban settlement, woody-fuelled fires cause considerable damage to life and property and heavily dominate politics, policy, and perceptions of vegetation fires. Most of these fires are in intrinsically flammable, open ecosystems with a long evolutionary history of fire. Closed forests may also burn, especially if subjected to logging and deforestation. Even relatively intact closed tropical forests burn following severe drought and/ or if logging has altered the structure, creating more ground fuels (Cochrane 2003; 2009). These forest fires, though small in extent at global scales, yield most of the greenhouse gases that persist in the atmosphere. Grass-fuelled fires, though very extensive, rapidly take up gases released by burning as post-burn grass productivity is high (van der Werf et al. 2010). Fires that burn woody fuels result in higher greenhouse gas emissions and take longer to recover from burns.

The vast extent of savanna fires, especially in Africa, bias global statistics on fire occurrence. For example, a 'general' pattern of fire activity in relation to NPP has been described with low fire activity at low NPP, highest activity at intermediate NPP and low fire activity at high NPP. The pattern has been explained as too little fuel to burn at low NPP, and too wet to burn at high NPP (in tropical forests). However, Archibald et al. (2018) pointed out that the pattern was essentially that of grass-fuelled savanna fires in Africa, which dominate global fire patterns. The relationship breaks down if fire patterns are studied on a biome basis.

7.4 Patterns of global fires: pyromes

Satellites have revealed the vast extent of wildfires. Fire is, indeed, a consumer of global significance. However, the extent to which fires shape ecosystems (and ecosystems shape fires) varies greatly depending on the fire regime (Box 7.1). Archibald et al. (2013) have mapped global fire regimes based on five fire variables measured from satellites. These are the size, frequency, intensity, season length, and spatial extent of individual fires (Box 7.1). They found five clusters of similar patterns of the fire variables which they called 'pyromes'. Unfortunately, satellites do not measure fire 'type', whether fires are surface, crown, mixed, or ground fires. These are easily identified on the ground and have been one of the most useful qualitative fire variables for predicting plant traits for fire survival (e.g. Pausas et al. 2004 for crown fires, Charles-Dominique et al. 2015b, 2018 for surface grass-fuelled fires).

7.4.1 Convergence and divergence of pyromes

You might expect fire regimes (pyromes) to converge with the same regime in similar climates and different regimes in different climates. A striking example of similar climates but different fire regimes has been recognized in the boreal regions of North America and Eurasia (Wirth 2005). Though the climates are similar, North American boreal forest fires occur at century+ intervals, are severe crown fires burning at high intensity, killing the trees, and have severe impacts on soils. They contrast strikingly with Eurasian boreal forest fires such as those in Siberia, which are surface fires, burning at intervals of 30 to 50 years at low intensities, and which the dominant trees usually survive (Figure 7.3; Plate 14). The North American boreal region is dominated by black spruce, *Picea mariana*, a serotinous conifer, which retains lower branches, acting as fuel ladders. Fire climbs the ladders and burns into the canopy. These crown fires kill the trees which regenerate from seeds released in the serotinous cones. In striking contrast, the Eurasian boreal forests are dominated by *Pinus sylvestris* and *Larix* spp with thick bark, self-pruning, with no fuel ladders. The trees survive surface fires burning through the understorey. These continental differences in fire regimes, despite similar boreal climates, are attributed to the different mix of raw materials, the tree species, and their attributes that fuel the fires (Rogers et al. 2015, Wirth 2005).

The opposite case, of very similar fire regimes under very different climates, occurs in the longleaf pine savannas of North Carolina, which have strikingly similar fire regimes to eucalypt savannas in the tropical climate near Darwin in Northern Australia. Both grassy ecosystems support frequent grass and litter-fuelled surface fires with a tall tree layer that is largely unaffected by fires. The eucalypt savannas occur under a monsoonal climate with a long (eight-month) dry season contrasting with the non-seasonal rainfall in North Carolina (Archibald et al. 2018). The North Carolina savanna is particularly puzzling since savannas are generally associated with a long dry season (e.g. Lehmann et al. 2011) but the rainfall of North Carolina is non-seasonal.

Are flammable systems at their climate/soil potential?

Where flammable ecosystems are widespread, they can easily be viewed as the 'zonal' vegetation, the physiological optimum mix of plants for the climate and soils. A good example is the flammable heathlands of Australia and South Africa. The heathlands are commonly assumed to be well adapted to the

Figure 7.3 In this mature jack pine stand (*Pinus banksiana*) the surface fire on the left is transitioning to a crown fire on the right. Jack pine is killed by crown fires but recovers from seeds stored in serotinous cones. Fire near Fort Providence, Canadian North-West Territories. With thanks to Stefan Doerr. (See Plate 14.)

abiotic environment of extremely nutrient-poor soils, even though the climate is warm enough and wet enough to support forests (Chapter 1). However, though pyrophilic heathlands are the dominant vegetation, closed forests do occur on soils derived from the same extremely pure quartzitic rocks (e.g. Cramer et al. 2018). The presence of closed forest communities in landscapes otherwise dominated by flammable shrublands cannot be overlooked in understanding determinants of these heathlands. Bowman (2000) analysed the distribution of 'rain-forests' in Australia. These are closed woody forests and thickets which occur as small, sometimes tiny, patches in a vast sea of flammable pyrophilic sclerophyll communities in Australia. Bowman explored all the usual abiotic explanations for their existence—local microclimate, soils with richer

nutrients, anthropogenic reduction of forest by fires etc.—but, ultimately, found exceptions to all of them and was left with major consumer control of most of the continent by recurrent fires as the most plausible general explanation (see also Orians & Milewski 2007).

Open ecosystems are generally low biomass systems relative to closed ecosystems. Australian vegetation provides spectacular exceptions. Eucalypts tower above the flammable understorey, even where, as in Tasmania, the understorey is itself a woodland. Yet despite the enormous biomass in tall eucalypt 'forests', they are remarkably open and transmit enough light to support an extremely shade-intolerant heathland understorey (Figure 7.4). After long periods of fire suppression, shade-tolerant species colonize the understorey of some eucalypt

Figure 7.4 Eucalypts are the most fire-tolerant trees, able to grow above highly flammable shrubby understoreys. a: Eucalypts have very thin canopies with vertically hanging leaves that transmit high light; b: consequently they support shade-intolerant flammable understorey plants including trees, shrubs, and grasses; c: they survive fires in the understorey by resprouting from d: deeply buried, well insulated buds and meristematic strands.

'forests' where they cast sufficient shade to prevent sun-loving eucalypts from recruiting. Without a consumer, fire, to open the vegetation, eucalypts would eventually be displaced by densely shading trees and shrubs. Eucalypts are probably globally unique in their ability to grow as trees above highly flammable understorey shrublands which burn with high intensity. Their ability to do so depends both on remarkably well-insulated buds and extremely rapid growth rates (Box 7.2).

Box 7.2 The remarkable eucalypts

Eucalypts are quintessentially Australian trees. Eucalypts (*Eucalyptus* and *Corymbia*, sensu lato) together account for most trees in most vegetation types on the Australian continent. No other continent has such complete dominance by such closely related species. Eucalypts are nearly ubiquitous in Australian ecosystems, dominating communities from savannas in the tropics, to mallee shrublands in the dry interior, to the enormous forests, with the tallest angiosperm trees in the world, in the humid temperate climate of eastern Australia and Tasmania. Eucalypts have a remarkable tolerance of high-intensity fires. They are the only trees that regularly overtop highly flammable understorey communities. Other tree taxa take too long to grow above the lethal flame zone created by high-biomass flammable shrublands and fail to emerge into the overstorey. An important clue to the eucalypt success story was revealed by Burrows (2002) who reported the remarkable insulation of regenerative tissues, buds and meristematic strands. They are buried under thick eucalypt bark or, sometimes, below the bark in the wood. These buds survive fires and allow the canopies to recover rapidly from intense fires. The second contribution to eucalypt success is surely their remarkable capacity for rapid stem growth, a feature much exploited by foresters around the world. These rapid growth rates allow eucalypts to grow above the flame zones, between fires, across a broad range of fire regimes.

The uniqueness of eucalypts is hard to recognize because where they are common there are typically no other tree lineages. The savannas of the Northern Territory are an exception where they share the overstorey with other non-eucalypt tree species, allowing *in situ* comparison of plant traits. Savanna fire intensities and flame heights are lower than those of woody-fuelled fires and many more tree species are able to escape the flame zone. Well-insulated buds are a common feature of most savanna trees, not only eucalypts, including many African savanna species (Charles-Dominique et al. 2015a). Rapid canopy recovery from epicormic sprouting is common in savanna trees. However, what remains distinctive is the very fast growth rate of eucalypts. Fensham and Bowman (1992) reported saplings growing to a height of 8m in 3 years in a savanna woodland gap (Fensham & Bowman 1992). Rapid height growth should allow more frequent escape from the flame zone in eucalypts versus non-eucalypts. A six-year fire experiment, in which hundreds of trees were labelled in different treatments, and followed for a decade, confirmed that this was the case (Bond et al. 2012). Six times more eucalypts escaped the fire trap (growing taller than 4m) than non-eucalypt saplings (Figure 1). This difference in escape rate was similar to the proportions of trees in the woodland canopy where eucalypts outnumbered non-eucalypts by at least six to one.

Studies of carbon gain per unit leaf mass have shown that the eucalypt species in these savannas produce from 1.5 to seven times more carbon per gram of leaf than non-eucalypts in the same community! The basis for this remarkable bioengineering feat (imagine driving seven times further for the same amount of fuel) is not known. Leaf photosynthetic rates are unremarkable in eucalypts compared to other savanna trees and leaf lifespans are also equivalent (Eamus et al. 1999). Where eucalypts do differ is in the vertically oriented leaves and very open, sparsely-branched canopies (King 1997; Figure 2). The low leaf area minimizes self-shading in eucalypt canopies, contrasting strikingly with forest trees. Are eucalypts so successful in open ecosystems because they are opting for high carbon gain? Rapid height growth is a major selective advantage in open ecosystems where trees have to escape the fire trap. In forests, where trees compete for light, exclusion of neighbours by producing a dense, self-shading canopy, which would eliminate eucalypts with their open, shade-intolerant canopies, seems a requirement so the open eucalypt canopy would be a major disadvantage. Regardless of the mechanism, eucalypts are major outliers in open ecosystems globally (Lehmann et al. 2014; Moncrieff et al. 2014c). Though they form high biomass 'forests', these are not 'closed' forests. Beneath the sparsely leaved canopies are understoreys of highly flammable shade-intolerant woody plants, ferns, and grasses. Contrary to the general pattern of low biomass in open ecosystems, Australian open ecosystems include the tallest trees, and most carbon-rich communities in the world (Tng et al. 2012).

continued

Box 7.2 *Continued*

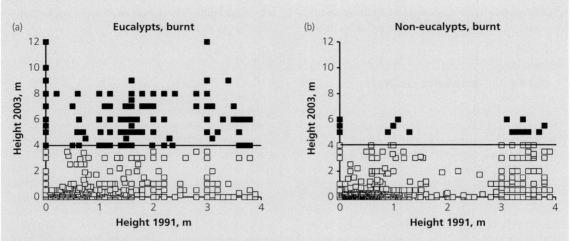

Figure 1. Changes in juvenile tree height from 1991 to 2003 for eucalypts and non-eucalypts in the Kapalga savanna fire experiment. Solid points are plants that grew to >4 m tall by 2003. Above this height the risk of topkill is reduced. 127 of 706 tagged eucalypts and 19 of 934 tagged non-eucalypts escaped over the period. From Bond et al. (2012).

Figure 2. Eucalypts (right) are strikingly different from many other trees (left). Many species have vertically oriented leaves and a very sparse canopy yet they attain very rapid growth rates. Forest trees, in contrast, have dense canopies and experience considerable self-shading in the competitive forest environment.

7.5 Evidence for fire-maintained open ecosystems

7.5.1 Patterns

The existence of recurrent fires and a fire regime is necessary but not sufficient evidence that fires maintain open ecosystems. In Australia, the most flammable continent, debates on the importance of fire have continued for decades without resolution.

The potential for a region to support tall trees in a closed forest can now be explored at diverse scales with DGVMs (Chapter 2). These process-based simulations indicate that vast areas of the world could support much more forest (>40 per cent increase in forest area) if fires could be 'switched off'. Empirical studies have explored tree abundance over soil moisture gradients in the tropics. For example, Sankaran et al. (2005) assembled a large data set of tree cover in Africa suggesting climate limits to tree cover below 650 mm MAP and disturbance (consumer) limits to tree cover above 650 mm. Lehmann et al. (2011) analysed vegetation maps exploring open versus closed ecosystems in Australia, South America, and Africa (Chapter 2, Figure 2.6). They showed that open savannas closely corresponded with frequent fire, except in Africa where extensive savannas occur in drier climates with few or no fires. A key finding of this study was that, across all rainfall zones, open ecosystems never completely dominated but there was always some fraction of closed woody vegetation with the latter never falling below 20 per cent of the area sampled. Thus mosaics of closed forest and flammable open grassy ecosystems are the norm over much of these three continents.

7.5.2 Manipulative studies

If fire maintains open vegetation, then suppressing fires should lead to their conversion to closed forests and thickets. The reverse process, adding fire to a closed system, should potentially open up forest and replace it with an open pyrophilic system. Evaluating studies that have suppressed or promoted fire is complicated by the diverse metrics of vegetation change. By far the most common variables are various measures of the importance of

trees, including per cent cover, basal area, stem density, stem size classes, and compositional shifts. Changes in the understorey, to the herbaceous layer, are often ignored although they are most revealing for the closed/open dichotomy (Chapter 1). If the understorey has a high and relatively continuous cover of shade-intolerant herbs or shrubs, the system is 'open' and will readily burn. If the understorey lacks shade-intolerant plants, it is likely 'closed' and will lack the fuel to support rapidly spreading fires. Unfortunately, vegetation classification is commonly based on trees, rather than the understorey. The broad definition of 'forest' used by the FAO completely confounds open and closed vegetation and therefore completely misses the distinction in processes between these two categories (Parr et al. 2014; Veldman et al. 2015a). The problem is very apparent in south and south-east Asia where the word 'forest' is used for both open, pyrophilic ecosystems and closed, pyrophobic forests. The consequence of muddled terminology has been that fires have been suppressed in teak (*Tectona grandis*) 'forests', which are pyrophilic, resulting in invasion of non-native weeds, such as *Lantana* and *Chromolaena*, which suppress recruitment of teak and other savanna tree species (Figure 7.5).

7.5.3 Fire suppression in tropical and subtropical grassy ecosystems

Experiments

There are many experiments manipulating fire scattered in different ecosystems of the world. They vary greatly in scale, variables measured, replication, and presence of controls. Some have been maintained for many decades, defying critiques of statistical design, utility, relevance to current management practices, etc. Some of the shortest experiments have yielded numerous papers (e.g. Kakadu in northern Australia, a 6-year experiment; Andersen et al. 2003) while some of the longest running ones have yielded hardly any (e.g. the West African experiments; Laris & Wardell 2006). But regardless of these various design flaws, they are invaluable in demonstrating: 1) rates of change; 2) direction and magnitude of change; 3) compositional shifts

Figure 7.5 Top: Fire-protected mesic savanna ('dry forest') in India with heavy infestation of non-native *Chromolaena* and *Lantana* weeds. Bottom: Frequently burnt savanna in the same landscape with few invasive plants and dominated by the native C_4 grassy understorey.

towards or away from forests; 4) fire manipulation as a potential cause of trophic cascades in species composition.

There are, and were, many field experiments exploring different burning treatments. They were usually set up to inform the use of fire in managing vegetation. Few would pass muster for modern experimental design. For example, many lack quantitative descriptions of initial conditions, including felling of all trees as one practice, are poorly replicated, or lack follow-up measurements of key variables. Nevertheless they have been maintained for decades and are extremely useful in indicating where fire matters in determining ecosystem structure and function.

Cedar Creek in Minnesota is famous for experiments exploring biodiversity, stability, and ecosystem

function (Tilman et al. 2006). The vegetation is an oak savanna (bur oak, *Quercus macrocarpa*, and pin oak, *Q. ellipsoidalis*) with a herb layer of C_4 andropogonoid grasses maintained by frequent fires. Long-term fire experiments have shown that ecosystem structure and function depend on frequent fires. Fire suppression for 30 years converted the oak savanna to a closed forest dominated by pin oak but with increasing densities of red maple (*Acer rubrum*), *Prunus serotina*, and *Amelanchier* sp. and the loss of the shade-intolerant C_4 grasses (Figure 7.6; Peterson & Reich 2001, 2008). Bur oak, the savanna dominant, failed to recruit in the shady forest understorey. Ironically, these burning experiments indicate that it is not biodiversity that stabilizes the herbaceous ecosystem, for which Cedar Creek is famous, but frequent fires (cf. Tilman et al. 2006).

Similar experiments have been conducted for decades in the longleaf pine (*Pinus palustris*) savannas of the southern USA. Here too, fire suppression has led to the loss of the C_4 grass layer dominated by *Aristida*. The pine savanna was replaced by a closed broadleaved oak–maple forest. The rate of closure varied with slope position and soil type and was slowest on bottomland locations. Though longleaf pine savannas have long been considered anthropogenic artefacts, they are now recognized as ancient open ecosystems (Noss 2012; Noss et al. 2015).

Fire suppression experiments have led to similar results of increased tree densities, changes in vegetation structure, tree species composition, and, where reported, changes in grass species to less flammable lineages or loss of grasses altogether. An example from the Ivory Coast in Africa, set up by Aubreville in the 1930s, is shown in Figure 7.7 (Louppe et al. 1995). This design of early dry season fires (low intensity), late dry season fires (high intensity), and no fires was repeated in several African countries with similar results. Early dry season fires show an increase in trees and a shift in tree species composition to more fire-sensitive (forest) species. Late dry season burning treatments show a decrease in trees and reduced tree diversity. Fire suppression results in increases in tree densities and size of existing saplings, a shift in tree species composition towards closed forest, and loss of the

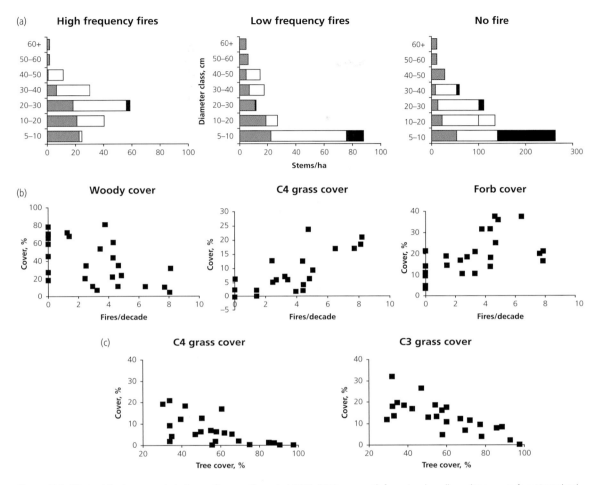

Figure 7.6 Effects of fire frequency, including no fire, over the period 1962–1995 on growth forms in a broadleaved temperate forest/grassland mosaic, Minnesota, USA. A. Woody structure and composition in response to high frequency (11–26 fires), low frequency (4 fires), and no fires on tree structure and composition; grey = bur oak, white = pin oak, black = other, mostly closed forest species. B. Response of woody, C_4 grassy and forb cover to fire frequency. C: Response of grasses to shading (tree cover) by woody plants. Data from Peterson & Reich (2001, 2008).

grass layer. Though shade-tolerant species colonize these treatments, the loss of savanna trees is slow. There is often an interaction with soils such that changes in woody strata are fastest on deep, well-drained soils and slowest where soils are shallow, or poorly drained on lower slopes. Nevertheless, over the long term, forests may also spread to colonize these less favourable soils (Louppe et al. 1995). In Venezuela, a long-term fire suppression experiment did not result in canopy closure despite decades of fire suppression. This was attributed to the shallow soils over ironpan impeding forest development (San José & Fariñas 1983).

Fire experiments in semi-arid savannas have been maintained since the 1950s at Kruger National Park in South Africa. This is a large replicated experiment set up in four different community types with different geologies and rainfall. Treatments vary in frequency and season of burn and include a fire suppression treatment. Fire suppression has not led to development of closed woody communities except at the highest rainfall sites (700–750 mm) on granite-derived sandy soils (Higgins et al. 2007; Smit et al. 2010). Bond et al. (2003a) collated qualitative trends of numerous fire experiments in South Africa and concluded that trends to forest with fire

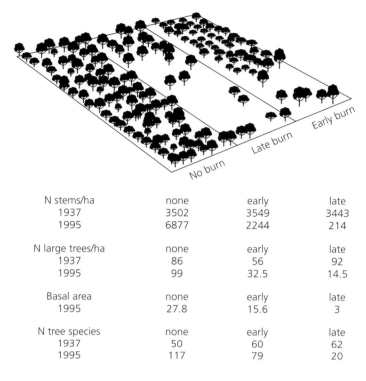

N stems/ha	none	early	late
1937	3502	3549	3443
1995	6877	2244	214

N large trees/ha	none	early	late
1937	86	56	92
1995	99	32.5	14.5

Basal area	none	early	late
1995	27.8	15.6	3

N tree species	none	early	late
1937	50	60	62
1995	117	79	20

Figure 7.7 Aubreville fire plots, layout, and results. The three replicates, each 100 × 200m, were laid out along a slope with different soils at top and bottom slope. The bottom slopes remained open the longest but had closed over to form forest in the no burn treatment after ~60 years of fire exclusion. Data from Louppe et al. (1995).

suppression were only apparent in higher rainfall sites (>700 mm MAP). The lack of a trend to compositional shifts to forest in the drier Kruger sites is consistent with this general pattern in subtropical Africa.

Landscape-scale shifts

Fire suppression has been deliberately imposed by legislation in some countries, with varying degrees of success. Suppression has also occurred as a consequence of habitat fragmentation by roads, cultivated fields, and settlements. Deliberate fire suppression was practised for a century and more in the USA. The result has been near elimination of the grasslands and savannas of the south and their conversion to closed broadleaved forest. Noss (2012) has described these 'forgotten grasslands' arguing that their remarkable diversity represents an unrecognized global biodiversity hotspot (Noss et al. 2015). Far from being of anthropogenic origin, these are ancient fire-maintained systems very rich in endemic species (Chapter 4). A similar fate occurred in oak savannas in the eastern USA, leading to what

Nowacki and Abrams (2008) have described as the 'mesophication' of the eastern USA. Grassland fires are relatively easily managed and suppressed, and removal of fire has led to their widespread conversion to broadleaved forests in the region. Thus the trends seen in the Cedar Creek experiments are replicated at much larger scales in the eastern USA.

In Brazil, the savannas, the 'cerrado', have been severely fragmented by land conversion for crop farming. In the small savanna remnants, fires were reduced and actively suppressed in response to national legislation prohibiting fires. The consequence has been extensive conversion of open cerrado to closed woody formations with greatly reduced or no grass cover (Durigan & Ratter 2006; Pinheiro & Durigan 2009). Unlike Amazon forests, conversion of cerrado to soya beans, maize, and pasture grasses has proceeded with little national or international opposition until very recently. Reintroduction of fire, requiring new training and novel legislation, will be required to restore this ecological process to the cerrado (Durigan & Ratter 2016).

In Gabon, in central Africa, forests are invading savannas in protected areas. The savannas are valued as habitat, including edge habitat, for forest elephants (*Loxodonta cyclotis*) and attempts are being made to reintroduce fire and to develop burning techniques that maximize fire intensity to help stop forest spread (Jeffery et al. 2014). In South Africa, savannas have been colonized by a dry forest ('thicket') leading to a biome switch from open to closed vegetation with loss of grasses, loss of shade-intolerant forb species, the slow death of savanna trees, and turnover of ant species to forest dwellers (Parr et al. 2012). Forbs in the grassy biomes in this system are perennial with large underground storage organs (Chapter 4), very resilient to frequent fires but extremely intolerant of shade. Once eliminated by shading from invading thicket, they may take many decades (centuries?) to recolonize secondary grasslands (Zaloumis & Bond 2016).

In northern Australia, Ondei et al. (2017) analysed rainforest areas over 55 years from 1949 to 2005. They compared two areas sharing the same climate and geology but with different consumer pressure. The Mitchell plateau study site had frequent fires (average of 0.58 fires per year) and cattle were common while the Bougainville peninsula had rare fires (0.11 per year) and no livestock. Rainforests expanded more (69 per cent) where fires were rare versus where they were common (9 per cent) (Figure 7.8). Topography influenced forest expansion with the least expansion on level terrain and then only on Bougainville where fires were rare. Rainforest expansion can be partly attributed to increasing rainfall over the half century at both sites but the expansion at a landscape scale is modified by fire activity. Forest margins were typically abrupt where fires were frequent but much more mixed where fires were rare. This result of sharp boundaries with major species turnover where fires are frequent, and blended savanna and forest elements where fires have been suppressed is common throughout the tropics. Cerradão, the most densely wooded formation under the umbrella of 'cerrado', is likely also a savanna in transition to a forest with low grass cover and a general lack of savanna tree recruitment in the cerradão woodland contrasting strikingly with frequent forest tree recruits (Franczak et al. 2011).

(a)
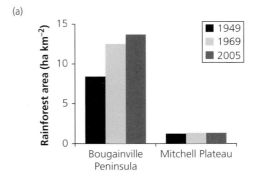

(b)
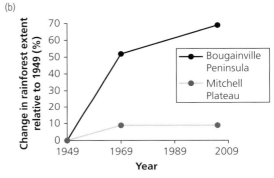

Figure 7.8 Change in rainforest distribution in two matched sites in the Kimberley range, north-west Australia, 1949–2009. Fires were suppressed on Bougainville Peninsula over the study period whereas the Mitchell Plateau burned frequently and also had livestock present. The two sites share the same geology and similar climate. From Ondei et al. (2017).

This brief survey of fire exclusion studies indicates that fires are essential for the maintenance of most open grassy systems in more productive sites. Where fires are suppressed for several decades there is a complete biome shift to closed forest with cascading changes in composition. The rates of change appear to be faster in more mesic climates and more fertile soils. But, given sufficient time, soils and sites that were slow to change may also convert to forest.

7.5.4 Fire suppression in woody and grassy temperate ecosystems

There seem to be few published reports on manipulation of fire regimes in Eurasian steppe/forest transition areas. And there are very few manipulative studies of fire frequency, including fire suppression, in woody-fuelled vegetation, including Mediterranean-type shrublands, heathlands, and conifer and eucalypt woodlands. The pace of change is apparently far slower in woody-fuelled systems. Jackson's concept of fire-driven assembly of Tasmanian communities described fire frequencies measured in 500 years +. Modification of soil nutrients was among the major feedbacks inferred by Jackson as necessary to maintain the different communities. Such soil changes have been observed following forest expansion into savanna (e.g. Silva et al. 2008, 2013; Louppe et al. 1995) over periods of a decade or more or even following the emergence of a tree from the grass layer in savannas (Ludwig et al. 2004; Coetsee et al. 2010; Chapter 6).

Poulsen and Hoffman (2015) used aerial photos (1944 to 2008) and repeat landscape photos to determine the effects of long-term fire exclusion on forest replacement of fynbos on Table Mountain in the Cape region of South Africa. The general trend from 1944 to 2008 was one of increasing forest reflected in changes in the number of forest patches (149 to 174) and a 65 per cent increase in forest area from 884 ha to 1461 ha. Only 13 of the forest patches decreased in size, 65 were stable and 96 increased in size. The study distinguished between mesic Afrotemperate forest and the drier milkwood forest (*Sideroxylon inerme*) with the former expanding the most. The authors argued that reduction in the use of fire from the twentieth century, especially relative to the nineteenth century, was largely responsible for forest increase.

7.6 Evidence for fire-maintained open ecosystems: adding fire

Can forests be converted to open ecosystems by adding fire? Here I do not mean clearing forests for agriculture but, rather, spreading fire from a flammable open ecosystem into a closed forest. In the tropics, this process has been labelled 'savannization' and is one of the ominous futures projected for Amazon forests. Reports of fires in the Amazon emerged during the 1997–1998 El Niño, a hot dry period (Nepstad et al. 1999). Over the same period, tropical forests in Indonesia suffered severe fires and smoke pollution was unprecedented. The story of the devastation wreaked by fire in these forests is reported so often in

the popular media that it hardly seems worth asking whether adding fire can open up forests. Yet the conundrum is that forest patches are common and widespread in the tropics adjoining highly flammable savannas. If forests are so vulnerable to burning, why do these forest patches persist? The same could be asked of any of the boundaries between pyrophilic and pyrophobic vegetation and is discussed further below in section 7.8.

Formal fire addition experiments have been set up in the Amazon forests. One such experiment is in the drier transitional forests towards their boundary with savannas. Balch et al. (2015) recorded grass cover in controls versus sites burned every three years or burned repeatedly on an annual basis. The infrequent fires in their study site had larger effects on tree mortality and grass invasion than the annual burns as there was more fuel to burn. Thus aboveground biomass was reduced to one third of the control in the annual burn and to less than 10 per cent of the control in the triennial burn. Grass biomass at the edge exceeded 10 Mgha^{-1}, a very heavy fuel load for grass fires. In South Africa, forests recovered from a single fire and suppressed grass growth in less than a decade after a severe fire. In the South African situation, successive fires at intervals of two to four years succeeded in converting the burnt forests to grasslands (Beckett 2018).

Most studies of addition of fires to closed ecosystems are associated with the introduction of flammable invasive species (Brooks et al. 2004). A classic example is the invasion of Hawaiian forests by C_4 grasses. The grasses are highly flammable and initiated a grass-fire cycle that converted the native forest to secondary grassland over the seasonally dry climate regions of Hawaii (D'Antonio & Vitousek 1992; Tunison et al. 2000; D'Antonio et al. 2011). The humid non-seasonal rainfall areas facing the prevailing winds have not been converted to pyrophilic grasslands. Follow-up studies have shown that repeated fires are not essential for maintaining the altered grassland state. Thus a combination of fire to kill the native species and, probably, grass competition to exclude subsequent recolonization, is enough to transform fire-naïve island forests to an open grassland.

Madagascar was thought to be an example of a forested system transformed by human settlement a mere two thousand years ago by a grass-fire cycle that transformed the island to a smouldering ruin of deforested and degraded grassland. However, this narrative has been questioned and there is growing evidence that the fires, the grasslands, and the open habitat biota, including the grasses, are ancient (Chapter 4; Bond et al. 2008; Burney 1987a,b; Vorontsova et al. 2016; Hackel et al. 2018). Though it is highly likely that the grasslands expanded their area aided by human-set fires, the original open grassland (and *Erica* heathland) was widespread across the island and the extent to which forests retreated with burning in the first millennia after human settlement has yet to be determined.

7.7 Alternative stable states

7.7.1 Pyrophilic versus pyrophobic ecosystems

Fire-promoting versus fire-resisting ecosystems were among the first to be identified as possible terrestrial examples of real-life alternative stable states (ASS). Wilson and Agnew (1992) noted how flammability could maintain alternative fire-promoting and fire-suppressing vegetation states. The dichotomy has since been proposed as an example of ASS for many vegetation types in diverse regions including pine savannas/hardwood forests in north Carolina (Beckage & Ellingwood 2008), eucalypt woodland versus closed rainforest in Queensland, Australia (Warman & Moles 2009), conifer forest versus chaparral in the western USA (Odion et al. 2010), and the complex mosaics of forests, woodlands, and shrublands in Tasmania (Wood & Bowman 2012) and fynbos and forest in South Africa (Coetsee et al. 2015, Cramer et al. 2018). The ecological importance of ASS is that it implies stability of very different ecosystems growing under the same environmental condition: open ecosystems are not early successional stages in transition to closed forests but stable systems that persist for millennia. From an evolutionary perspective, it makes sense that organisms evolve to maintain their habitat rather than losing it by facilitating invasion of later successional species. ASS could lead to the evolution of very different biota, with different functional traits and little or no overlap between alternative states. However, ASS theory also argues for the potential for rapid change to the alternative state. Instead of gradual successional change, ASS

carries the potential for 'catastrophic regime shifts' (Sheffer & Carpenter 2003). After such a shift, reversals are difficult, even if growing conditions also revert to the initial condition. ASS challenges traditional succession theory. It also challenges the idea that contrasting ecosystems are maintained by different resource availability and different abiotic settings. Instead, resource differences are seen as the consequence of biotic feedbacks of each state to the environment (see Chapter 3).

Two papers published back to back in *Science* in 2011 provided the first major challenge to traditional ideas on the assembly of tropical ecosystems (Staver et al. 2011a; Hirota et al. 2011). Both used MODIS satellite imagery to analyse patterns of tree cover across rainfall gradients in the tropics. You might expect trees to increase linearly along a rainfall gradient, eventually forming forests where wet enough. However the analyses showed that this was not the case; instead tree cover was multimodal with peaks at ~15–20 per cent and >80 per cent and troughs at 50–60 per cent. This pattern is consistent with ASS. Tree cover has two basins of attraction, one savanna, the other forest, with feedbacks, both positive and negative, producing bistable equilibrium basins.

Tree cover as used in these studies has some problems as a metric of tree importance (Hanan et al. 2014; Staver & Hanson 2015). Dantas et al. (2016) collated plot data on basal area, a proxy for biomass, from 100s of sites in Africa and South America. Unlike tree cover, which asymptotes at 100 per cent, basal area increases with resources and is a good predictor of Leaf Area Index (LAI) and therefore shade cast. As for the MODIS satellite dataset, they found peaks and troughs in tree importance. In Africa, there were three modes: a sparse wooded grassland, savannas (~7 m^2/ha), and forests (>20m^2/ha). South America had only two modes, lacking the grassland but with similar patterns for savannas and forests, except the continent has more high rainfall sites than Africa (Figure 7.9). The modal states shared similar environmental space characterized by climate variables (rainfall, rainfall seasonality, maximum and minimum temperature) soil texture, cation exchange capacity, and organic matter sampled to a depth of 80cm (Table 7.2). The three states differed

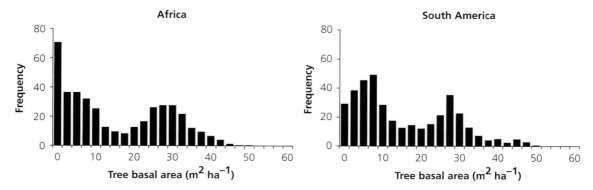

Figure 7.9 Tree basal area for tropical vegetation in Africa and South America. There are three modes in Africa—grassland, savanna, and forest—but only two in South America which lacks a grassland mode. Modified from Dantas et al. (2016).

Table 7.2 Resource space (rows) shared with other biomes (columns) in the Afrotropics and Neotropics. WG: wooded grassland state; S: savanna state; F: forest state; Both: both biome states. There is extensive resource overlap between forests and savannas on both continents. Wooded grassland, is restricted to Africa but overlaps in resource space with savannas and forests. From Dantas et al. (2016).

	Afrotropics				Neotropics	
	WG	S	F	Both	S	F
WG	–	87%	35%	89%	–	–
S	52%	–	41%	74%	–	68%
F	19%	37%	–	37%	45%	–

in fire history with the most fire in savannas, few in forests, and the least in the African grasslands. The grassland state appears to be short-grass communities maintained by heavy grazing by the African megafauna or livestock. In the jargon of Bond (2005), these are 'brown world' systems dominated by herbivores contrasting with 'black world' savannas where fires are the prominent consumer and the 'green world' where vegetation is at its climate-limited forest potential. These distinct states of tree structure support the ASS predictions in terms of pattern and zones of attraction. Neither succession nor abiotic determinism offers explanations for the peaks and troughs in tree frequencies, nor how they change across different rainfall classes (Figure 7.9).

7.7.2 Feedbacks maintaining states: flammability

As noted by many authors, fires are a major driver of state divergence into flammable versus fire-resistant formations. Feedbacks to ecosystem flammability are produced by the architecture of the common growth forms and the microclimate created by the vegetation. Grasses are highly flammable if fully cured (Ripley et al. 2015; Simpson et al. 2016). The rate and timing of curing varies among grass lineages. Andropogoneae leaves dry out earliest in the dry season although, in theory, the clade has the most water-use efficient subtype of photosynthesis. Andropogonoids are also the first to be frosted in the subtropics so that their leaves dry out long before other clades. Members of the Andropogoneae dominate the most frequently burnt grassy biomes of the world. Several species, such as *Themeda triandra*, dominate frequently burnt stands. They are extremely intolerant of shade and, if not defoliated by fire or grazing, die from self-shading of accumulated dead leaf litter from previous growing seasons (Tilman & Wedin 1991; Everson et al. 1988). These widespread species are the ultimate 'fire-weed', promoting rapidly spreading grass fires.

Traits promoting flammability in woody plants include small leaves, dense ramification, the retention of dead leaves or branches, and, for trees, a fuel ladder into the canopy (Pausas et al. 2017; Burger & Bond 2015; Schwilk 2015; Bowman et al. 2014; Chapter 4). Van Wilgen et al. (1990) compared traits contributing to flammability in fynbos versus fire-excluding closed evergreen forest in South Africa. The forests had more biomass to burn but primarily organized into large-diameter organs. Fynbos had finer branches and dried out far more rapidly. Burger and Bond (2015) experimentally burned fynbos and forest margin species and found striking differences in flammability that help explain why fynbos fires usually burn out on forest margins (Figure 1.3).

For ecologists raised on the assumption that organisms are designed for optimal resource capture in a given environment, it is hard to imagine plant design for promoting flammability. Mutch (1970) first pointed out that the growth forms characteristic of flammable vegetation maintained by frequent fire may have evolved flammability to maintain their existence by burning out forest trees that might shade them out. The hypothesis was criticized for being group selectionist. However, subsequent work has shown that flammability could evolve by a form of kin selection (Bond & Midgley 1995). Furthermore, populations of *Ulex* in Europe show variation in flammability depending on their past history of burning and the genetic basis for these differences has been demonstrated (Pausas et al. 2012, 2017). Phylogenetic studies have also revealed the evolution of syndromes of traits indicating selection for high flammability associated with fire-stimulated reproduction by means of serotinous cones (Schwilk & Ackerly 2001). Thus, evolved feedbacks selecting for enhanced flammability are beginning to look plausible (Archibald et al. 2018). The topic remains controversial, with the alternative hypothesis being that flammability enhancing traits are incidental to, usually unspecified, physiological benefits.

7.7.3 Feedbacks maintaining states: soils

Besides contrasting flammability, important feedbacks have also been recognized between alternative states and soils. Nutrient-enriched soils will favour rapid growth of forest trees enabling them to spread out and colonize open, flammable vegetation. In contrast, fires can impoverish soils by volatilizing nitrogen, and promoting post-burn losses of ash and soils. Effects of fires on nutrients are most pronounced where vegetation is green when burned. As leaves senesce, a significant fraction is resorbed. Crown fires burning evergreen shrublands are therefore most likely to show fire-induced reduction of topsoil nutrients. Striking differences in nutrients under forests and flammable shrublands growing on soils derived from the same geology have been reported in some systems (Cramer et al. 2018; Coetsee et al. 2015). Accumulation of nutrients under forests has also been reported where forests have invaded savannas in Brazil on nutrient-poor soils (Silva et al. 2013) and in West African savannas colonized by forest (Louppe et al. 1995). In South African comparisons of forest versus savannas and grasslands, nutrient differences were relatively small, even where the forests had colonized the grasslands for more than a thousand years (Gray & Bond 2015).

Among the most striking examples of contrasting soil nutrients in open versus closed ecosystems is the contrast between fynbos and closed forest in the south-western Cape of South Africa. Fynbos soils are notorious as being, along with south-western Australia, the most nutrient-poor soils in the world. They are derived from kilometre thick layers of nearly pure quartzitic sandstones. Nevertheless, forests occur on quartzite-derived soils and, in geographically matched sites, accumulate much higher nutrients than fynbos (Chapter 1, Table 1.2). Forest trees grow poorly in fynbos soils (Manders & Richardson 1992) so that forest colonization is rare, even with very long intervals of many decades between fires. Fynbos, in contrast, is home to species with diverse strategies for coping with very different nutrients, including cluster roots of shrubs, and the reed-like restios (Lambers et al. 2008). For more information on soils and the role of vegetation in modifying nutrient availability, see Chapter 6.

7.7.4 Stability of states

Demonstration of stability has been seen as the greatest stumbling block for ASS theory (e.g.

Petraitis 2013). However, paleoecological studies are continually exploring novel proxies for past vegetation and are helping to establish the stability of contrasting systems. Carbon isotopes are particularly useful in tropical systems since there is a large difference in the isotopic signal of C_4 grass-derived soil carbon versus C_3 tree-derived carbon. The soil carbon can be dated to a greater or lesser degree of accuracy. Isotope-based studies have shown long-term stability on the order of 1000s of years for grasslands, contrary to successional assumptions that they are seral to forest.

ASS theory has the demanding requirement of predicting both stability *and* rapid change. Carbon isotopes have been useful for this purpose too. In the forests of the Hluhluwe conservation area in South Africa, the mosaic of forest and savanna has long been assumed to be caused by deforestation when Iron Age famers settled in the area nearly two thousand years ago. However, isotope analyses of soil carbon showed that the grasslands were ancient and forests the new invading vegetation (Figure 7.10). The switch from grassland to forest is dated at about 1500 BP, implying stability of both states (uninvaded

grasslands persist) but also the capacity for radical biome shifts (Gillson 2015). Moncrieff et al. (2014b) explored the problem of stability of closed and open states in a different way. They simulated the vegetation of Africa using a DGVM but with different initial conditions, grass-dominated versus forest-dominated. Palaeoecological studies have shown that grasslands were far more extensive at the end of the last glacial so the grassland simulation is closer to the 'initial condition' of early Holocene landcover in Africa. Simulation results were compared with actual vegetation. The initial grassy simulation was a close match to current vegetation on the continent. The forest initial conditions simulated far too much forest. Savanna grasslands have apparently resisted conversion to forests, despite suitable climates, throughout the 12 000 years of the Holocene. This striking demonstration of the importance of history, of priority effects, indicates the potential stability of pyrophilic systems maintained by fire.

This stability of savanna/forest mixes over thousands of years is unexpected and surprising given evidence for the devastating effects of fires in tropical forests, and particularly the Amazon (e.g.

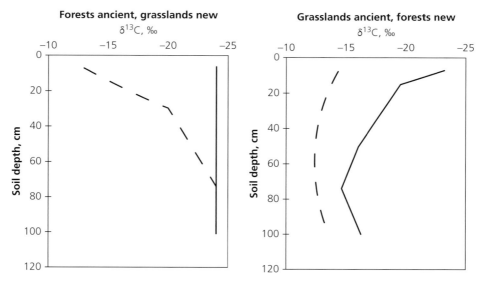

Figure 7.10 Stable isotopes of carbon can reveal past vegetation in tropical systems dominated by C_4 grasses. Deeper soil samples have older soil carbon so that depth is a surrogate for age. The figure on the left shows the expected pattern where forests are ancient (solid line), and grasslands are secondary (dashed line). The figure on the right shows the actual pattern in a South African forest/savanna mosaic where the grasslands (dashed) are ancient and have been colonized by forests (solid; ex Gray & Bond 2015). Examples such as these have forced revision of the idea that open systems are anthropogenically derived from forest.

Cochrane 2003, 2009; Barlow & Peres 2004; Balch et al. 2015). That Amazon forests exist at all seems a conundrum given their extreme sensitivity to burning. Why, on the one hand, are forests stable despite millennia of savannas burning while on the other, forests are apparently exceptionally vulnerable to a biome switch to grasslands after a few successive fires (e.g. Barlow & Peres 2004; Balch et al. 2015)? Selection for more fire-tolerant species near the forest edge is one possibility, implying that roads that penetrate to forest interiors could expose the most vulnerable trees to fire damage.

7.7.5 Regime shifts

Where fires continue to burn, boundaries between pyrophilic and pyrophobic ecosystems are surprisingly stable, given that fires are potentially so damaging to forest trees. Savanna fires generally peter out within a few metres of a hard forest edge (Figure 7.11). The same resistance to fire is common in shrubland/forest edges even though fires are generally more intense. Rare high-intensity fires, 'firestorms', usually driven by high wind speeds, do cross over stable boundaries and can burn deep

Figure 7.11 Fires from flammable open ecosystems seldom penetrate closed forests. A. Burnt fynbos shrublands on the left abutting unburnt forest to the right. B. Burnt grassland and *Erica* shrubs adjoining unburnt forest in Madagascar. Neither of the forest margins were protected from fires.

(several hundred metres) into the forest. There are very few examples in the literature, perhaps because the reviewers are reluctant to accept papers on single catastrophic events and want replication. Most studies of regime shifts, from forest to flammable system, follow deforestation where the structure of the forest has been modified by logging. Logging changes the microclimate and the distribution of 'fuel' creating more continuous fuel beds that facilitate fire spread.

Forest expansion into savanna

Incursion of forest trees into savannas, a slower paced regime shift, has been widely observed in many savanna landscapes over the last few decades. It has also been observed in many, but not all, long-term fire exclusion experiments. Satellite imagery and aerial photos have been used to study forest expansion into open ecosystems (e.g. Ondei et al. 2017; Warman & Moles 2009; Wigley et al. 2010; Silva et al. 2008; Nowacki & Abrams 2008).

Theory

Regime shifts have been explored theoretically in a growing number of studies (Staver et al. 2011b; Beckage & Ellingwood 2008; Beckage et al. 2009; van Nes et al. 2014). Beckage and Ellingwood (2008) simulated the longleaf pine system with three functional types; grass, pines (which grow with grass), and forest trees (broadleaved hardwood species). Using cellular automata they simulated fire spreading through the system. Cyclones also affect the system by causing tree mortality, opening up the system to more grass and promoting more fires. Their results showed that gradual climate change could cause abrupt ecological change through feedbacks between fire and vegetation.

Fires percolate through vegetation rather like liquid through ground coffee. Percolation theory as applied in landscape ecology predicts threshold conditions for fire spread. Archibald et al. (2009) explored the relationship between tree cover and fire spread. Using satellite-based remote sensing of tree cover and fire, they detected a threshold of tree cover at 40 per cent above which fires stopped burning (Figure 7.12). The prediction, then, is that if trees, or clumps of trees, inhibit

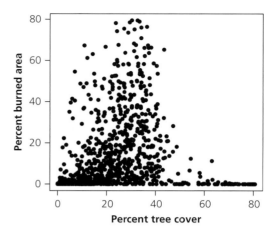

Figure 7.12 Percolation theory and fire. Burnt area in Africa south of the equator derived from MODIS burnt area product plotted against tree cover. Upper limits to burnt area decline steeply where tree cover, derived from MODIS products, exceeds 40%. From Archibald et al. (2009).

fire spread by suppressing grassy fuels, then fires will stop spreading once the trees reach 40 per cent cover of the land area. An important caveat is that the 40 per cent threshold was measured from satellite imagery and the relationship with ground-based estimates of foliage projected cover has yet to be determined.

The 40 per cent threshold for fire spread has also been observed at much smaller scales in a grazing experiment. Burning of the grazing treatments showed that fires failed to spread when grasses were shorter than 5cm height. Where short grass patches exceeded 40 per cent of the sample area, fires failed to spread through the area (Charles-Dominique et al. 2018). The application of percolation theory to fire suppression, or fire addition, at diverse scales looks promising for establishing threshold fuel conditions for fires to spread. At the landscape scale, the coalescence of patches of shade-casting trees and shrubs will not stop fires spreading until, according to the theory, they cover > 40 per cent of the area. Managers can use this threshold to guide prescribed fire interventions. A complication is that flammability of patches is not a constant but varies with fire intensity—extremely high-intensity fires will burn through almost any fuel configuration.

7.8 Life at the edge

The zone where two contrasting systems meet, the boundary between ASS, is clearly very important in understanding the mechanisms controlling persistence of fire-sensitive species in flammable vegetation or the conditions under which flammable vegetation might invade forests. Hoffmann et al. (2012a) developed a conceptual framework that elegantly summarizes the key considerations at the forest edge. Trees can either obtain relative

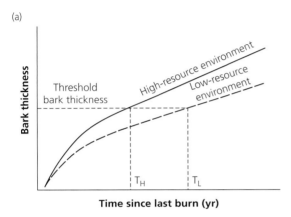

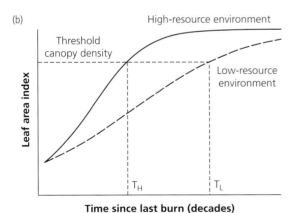

Figure 7.13 Two key thresholds influencing trees and the impact of fires, especially at forest margins. a) Fire resistance threshold is the size or bark thickness at which resprouting stems survive fires intact; b) fire suppression threshold is the canopy density needed to suppress flammable understorey cover, e.g. grasses. For both thresholds, the time taken to reach tipping points where either (a) trees survive fires or (b) fires do not burn for lack of fuel, varies with rates of growth and therefore with resources and intrinsic constraints on growth rates. Figure from Hoffmann et al. (2012a).

immunity from fire damage by growing to a size large enough to resist fire effects—the fire resistance threshold—or by growing to a size where they can suppress (shade out) the flammable understorey—the fire suppression threshold. The rate at which trees grow to resistance or suppression thresholds will influence the stability of the edge under a given fire regime. Site conditions will influence this rate, as will the intrinsic properties of the pool of tree species. The framework is illustrated in Figure 7.13. The data suggest that the resistance threshold is important for savanna trees to survive frequent fires but that the suppression threshold is seldom reached. This is because savanna trees have thin canopies transmitting enough light to support a grassy, shade-intolerant understorey (Chapter 4). Forest trees, in contrast, may seldom reach the resistance threshold because of thin bark, poorly insulated buds, or even hydraulic failure (Charles-Dominique et al. 2018). Instead, time to reach the suppression threshold is critical, determined by high LAI, and shading out flammable understorey growth forms . Studies in Brazil suggest that top-kill processes are the main determinant of forest/savanna species turnover at the forest edge. But observations in some African forest/savanna mosaics show that forest margin species recover from dormant seedbanks after fires penetrate the forest. Saplings growing from seed re-assemble the forest structure rapidly, providing shade that suppresses grass growth.

7.9 Summary

Fire is a very influential consumer and has been so for most of the history of terrestrial plant life. Fires were regularly burning vegetation for 100 million years before the first vertebrate herbivores evolved. What is puzzling is why it has taken researchers so long to study its role in shaping global vegetation. However, in the last decade, there has been remarkable progress in understanding the deep history of fire, the global reach of contemporary fires, areas sharing common fire regimes (pyromes), and the biogeography and ecology of pyrophilic and pyrophobic ecosystems. David Bowman invented the term 'pyrogeography' to describe this rapidly expanding global science of fire ecology (Bowman

& Wood 2009; Scott et al. 2014). Frequent grass-fuelled fires maintain tropical grassy biomes, often in mosaics with closed forests. These were among the first ecosystems to be considered alternative stable states with feedbacks to flammability in the grassy layer, or, in the forest, to microclimate and shade excluding fuels to support the fires. The multimodal distribution of tree cover and woody biomass discovered using MODIS imagery in savannas is consistent with alternative stable state theory. The implications are far-reaching and have attracted major research interest in savannas. That fire is the key agent excluding forest is supported by numerous examples of fire suppression resulting in increasing tree cover and forest expansion. But the scale of fire control and its climatic and edaphic context is not well established. The boreal forests are another major flammable open ecosystem, though whether they have the potential to change to closed forests, excluding light-loving understorey, is not well established. In the next chapter, we will see that fire and large herbivores are sometimes competing consumers driving divergent functional composition in ecosystems.

CHAPTER 8

Vertebrate herbivory and open ecosystems

'The general aspect of the island is now so barren and forbidding that some persons find it difficult to believe that it was once all green and fertile. This irreparable destruction was caused, in the first place, by goats. . . . These animals are the greatest of all foes to trees, because they eat off the young seedlings, and thus prevent the natural restoration of the forest.'

Alfred Russell Wallace (1892) on St Helena island

8.1 Introduction

Would a vegetation map of a park, a region, a continent, change if vertebrate herbivores went extinct? Many of the world's herbivores did go extinct a few thousand years ago. Did the extinctions trigger cascading consequences for ecosystems implying a fundamental role for mammals in maintaining open and closed ecosystems? Norman Owen-Smith (1987) first suggested that megaherbivores, animals of body size >1000 kg, played just such a role in maintaining open ecosystems in the Pleistocene. He argued that the end-Pleistocene extinction of the megafauna promoted forest expansion in the Holocene. In a similar vein, Zimov et al. (1995) in Russia, argued that the end-glacial transition from steppe grasslands to tundra was triggered, not by climate, but by extinction of the megafauna (animals >45 kg) with cascading consequences for the biota and the climate. Biotic change, rather than the glacial/interglacial climate transition was the major factor altering the vegetation according to these authors. In the 2000s, there has been an explosion of interest in the ecological impacts of the extinct megafauna (see e.g. Gill 2014; Johnson 2006; Bakker et al. 2016; Doughty et al. 2016a,b). The notion of 'rewilding', bringing large mammals back into ecosystems, has attracted wide scientific and popular interest. There have been serious attempts to recreate 'Pleistocene' parks, landscapes populated by a diversity of large mammal herbivores, and observing their ecosystem impacts. This chapter explores the importance of mammals as consumers creating or maintaining open ecosystems.

Africa retained a relatively intact megafauna and has provided important insights into Pleistocene ecology and the complex interactions of climate, large herbivores, fire, soils, and people. Africa's megafauna is now largely restricted to protected areas so that we can begin to observe how Africa's vegetation would change if it was to lose its remaining mammal species (Archibald & Hempson 2016; Hempson et al. 2017). Mammals have diverse impacts on ecosystems, altering nutrient cycling, dispersing fruits, feeding on seeds, browsing seedlings and saplings, consuming leaves, tearing off bark, trampling paths through dense vegetation, and more. The intention in this chapter is not to consider all these diverse activities but, rather, to consider whether mammal herbivory and mammal mechanical damage can account for open ecosystems where the climate potential is forest. How influential are mammal herbivores in regulating tree populations and tree species composition? The focus is not on the number of studies showing herbivore impacts but on those areas where herbivory is, was, or could be the major factor influencing vegetation.

Open Ecosystems: ecology and evolution beyond the forest edge. William J. Bond, Oxford University Press (2019). © William J. Bond 2019. DOI: 10.1093/oso/9780198812456.001.0001

8.2 Overgrazing, degradation, and other value-laden terms

Analyses of the potential of herbivores to alter ecosystems are laden with words reflecting implicit or explicit value judgments. Words such as 'overgrazing' and 'degraded' are very common in the literature but cloud the issue of assessing herbivore impacts. 'Overgrazing' is a term relating to a particular land use. Since cattle grazing is a common land use, 'overgrazing' is usually assessed in relation to cattle productivity. But heavy grazing that is bad for cattle might be excellent for white rhinoceros, warthogs, wildebeest, or prairie dogs! Degradation is also a value-laden term. The World Resources Institute produced a world map showing extent of 'degradation' (http://www.wri.org/resources/maps/atlas-forest-and-landscape-restoration-opportunities). The authors mapped the Serengeti and Kruger National Park, iconic savanna conservation areas in Africa, as 'degraded'. Their definition of 'degradation' is 'a process that reduces the volume and canopy cover of trees across a landscape' such as mammal herbivory or fire. 'Degradation' defined this way elevates trees above any other life form, plant or animal, and is clearly inappropriate for conserving African wildlife (Chapter 9). Deer 'overabundance' is another widely used term in temperate ecosystems also disguising values placed on different types of ecosystem (Côté et al. 2004). This chapter tries to avoid these terms in the hope of an open-minded consideration of whether, and how, animals can change vegetation structure by altering proportions of the major growth forms and their species composition.

8.3 Examples of herbivore impacts

Exclosures are by far the most common experimental approach for evaluating large vertebrate impacts on vegetation. The use of different types of fence for exclosures can help partition the effects among different sizes of herbivores (Veblen et al. 2016; van der Plas et al. 2016). The key response variables, from the perspective of open ecosystems, are tree abundance and composition, and changes in the understorey plants where these can feedback to the woody species. Many exclosure studies have been set up across diverse ecoregions of the world, providing potentially rich insights into herbivore impacts (Milchunas & Lauenroth 1993; Díaz et al. 2007; Tanentzap & Coomes 2012). Exclosure studies have been criticized for several shortcomings. They are often of too short duration relative to the lifespan of trees, biased to particular species or locations of management concern, and limited to the binary response of presence/absence of large herbivores (Bradshaw & Waller 2016). Experiments manipulating herbivore densities beyond just presence/absence require much higher levels of sustained management and are few and far between. Nevertheless, exclosures, like fire suppression studies, are invaluable for demonstrating qualitative changes from open to closed systems (Figure 8.1; Plate 15).

Figure 8.1 Effect of elk browsing in Yellowstone National Park. a) An aspen stand showing heavy bark damage from elk along the lower several metres of each tree and a long-term lack of recruitment with only large diameter trees present. b) High recruitment within a fenced exclosure. Root sprouts are present outside the exclosure but browsed short by elk. From Beschta & Ripple (2009). (See Plate 15.)

8.3.1 Exclosure experiments: temperate and boreal forests

Extensive networks of exclosures have been maintained for decades in North America and Eurasia. Deer numbers have grown to very high densities in many areas and provide an extreme example of whether mammal herbivores can alter ecosystem structure (Côté et al. 2004; Augustine & McNaughton 1998). Figure 8.2 shows one example from an exclosure study in the boreal forests of North America

investigating the impact of white-tailed deer (*Odocoileus virginianus*) (Hidding et al. 2013). The regional vegetation is fir–birch forest (*Abies balsamea, Betula papyrifera*). Deer were introduced to Anticosti Island (7943 km^2) in Quebec in 1896 and within 30 years had grown from the initial 200 animals to densities of >20 km^{-2} across the island. The fir and birch are being eliminated by the deer and replaced by spruce, *Picea* spp., which are the least palatable trees in the system. After logging, almost no trees recruited and those that did were exclusively *Picea*

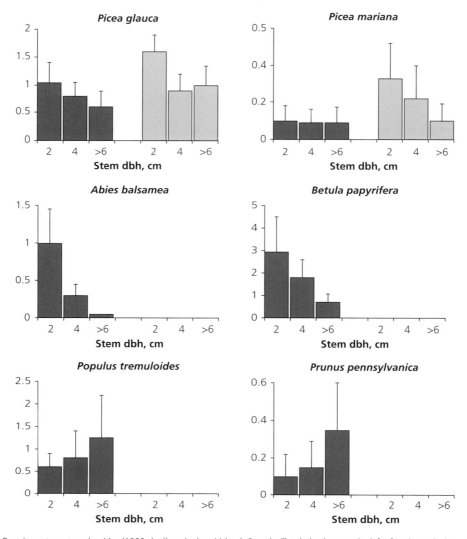

Figure 8.2 Deer impacts on tree densities (1000s ha^{-1}) on Anticosti island, Canada. The darker bars on the left of each graph show stem densities for three size classes protected by exclosures (H−). The light bars on the right show densities in the presence of deer (H+). The deciduous trees, and balsam fir, had been eliminated by deer and the system converted to a *Picea* 'fern savanna'. Redrawn from Hidding et al. (2013).

Figure 8.3 Much of the Scottish Highlands is open moorland but with the climate potential to support forests. Browsing by sheep and deer maintain the open habitat often in combination with acid soils that slow sapling growth. Photograph by William Bond.

spp. The deer is a mixed feeder (browser and grazer). It has opened up the shrub and tree layer, and compacted the soil by trampling. The ground layer has switched from forbs to light-demanding grasses and sedges. The original fir–birch forest has been transformed into an open ecosystem, a 'spruce savanna' (Hidding et al. 2013). Similar effects have been widely reported in the mid-west of the USA. Opening of the canopy also results in switches in the understorey to graminoids or to ferns such as *Dennstaedtia punctilobula* (Tanentzap et al. 2011). These aggressive understorey plants suppress tree regeneration, converting temperate broadleaved forest to a fern 'savanna'. A similar process has been observed in Japan where intense browsing by sika deer (> 50 deer km^{-2}) converted mixed deciduous woodland to grassland (dominated by *Zoysia japonica*). Reversion to forest only occurs with large-scale deer exclusion (Takatsuki 2009).

Browsing animals (deer, goats, sheep) seem particularly influential at higher latitudes or elevations. Scotland is renowned for its heathlands and moorlands yet the climate for most of the region is suitable

for trees. The moorlands were widely planted to conifers, particularly Sitka spruce (*Picea sitchensis*) showing that climate over most of the Scottish Highlands is quite capable of supporting conifer forests (Figure 8.3). Exclosure studies and regulation of deer and sheep densities suggest that the open system is in large part maintained by mammal browsing on slow-growing saplings (Palmer &Truscott 2003; Tanentzap et al. 2013). Fenton (2008) suggested that the Scottish uplands was the largest naturally open area left in Europe, kept open by a combination of poor soils, slow tree growth, and browsing by deer. The idea is contentious (e.g. Bennett 2009), challenging the idea of a Scottish primeval wildwood, but new developments in pollen analyses indicate that upland Scotland was much more open than mainland Europe throughout the Holocene (Fyfe et al. 2013).

8.3.2 Exclosure experiments: savannas

Exclosures have been widely employed in Africa to evaluate herbivore impacts on trees and grasses.

Results are complicated by simultaneous changes in the fire regime. Grasses protected from grazing grow tall and special measures must be taken to either burn within the exclosure site, risking damage to the infrastructure, or to exclude fire *and* herbivores. Figure 8.4 shows an example of results from an arid savanna in Kenya where fires have long been excluded from the landscape and herbivore impacts can be isolated (Sankaran et al. 2013). The exclosures were designed to keep out all herbivores from elephants to dikdik (*Madoqua* spp., <5kg antelope). A decade of exclusion resulted in many trees growing to escape height for antelope browsers (3m), a seven-fold increase in new recruits, and a three-fold increase in woody biomass. Thus, in this system, herbivores regulate woody cover, especially of the dominant acacia species. Sankaran and

colleagues note that removal of herbivores would result in rapid woody thickening transforming the ecosystem. In the same study area, with the same mammal fauna, but on different soils (black cracking clays), *Acacia drepanolobium* showed negligible effects of browsing (Sankaran et al. 2013). This species is extremely well defended by both spines and galls containing aggressive ants. Elephants avoid it (biting ants crawl up their trunk) as do antelope. Contingent effects such as these complicate generalizations about herbivore impacts.

Quantifying herbivore damage to trees has been transformed by the development of lidar, a powerful new tool for exploring consumer impacts on vegetation structure. Asner et al. (2009) and Asner and Levick (2012) used lidar very effectively to document megaherbivore (elephant) impacts on

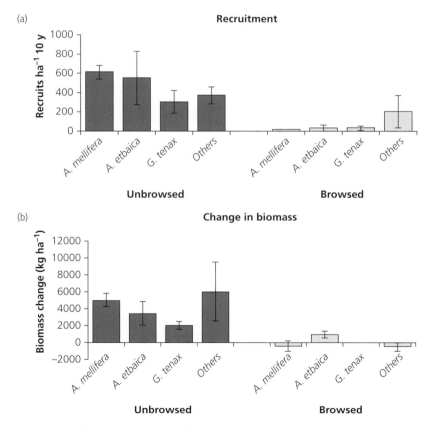

Figure 8.4 Decadal changes in an African savanna—protected from browsing and exposed to browsers: (a) shows browsing effects on density of new recruits and (b) changes in biomass (growth) of the common woody species in the area: *Acacia mellifera*, *A. etbaica*, *Grewia tenax*, and other species. Redrawn from Sankaran et al. (2013).

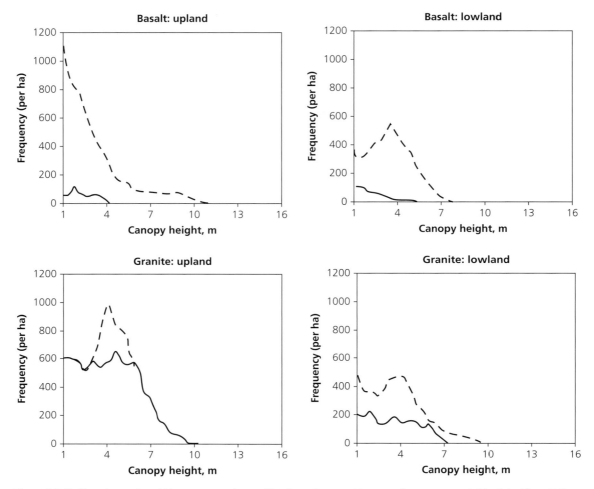

Figure 8.5 Herbivore impacts in an African savanna park vary with soil type. Frequency histogram of savanna canopy height derived from 1640 ha of airborne LIDAR observations. Dashed lines show vegetation protected from herbivory for >35 years; solid lines show vegetation exposed to herbivores. Elephants are major herbivores in this system. Redrawn from Asner et al. (2009). (See Plate 15.)

trees in the Kruger National Park (Figure 8.5). In ~40 years, elephants profoundly altered the structure of a savanna as reflected in different vegetation structure inside and outside a large (300 ha) exclosure (Figure 8.6; Plate 15). In the same study area, long-term fire exclusion had minor effects on tree sizes which, however, were all exposed to elephants. This study is the most convincing experimental study of which I am aware for demonstrating how elephants can impact a tree community. The environmental setting is nutrient-rich shallow basalt-derived clays in a semi-arid climate (MAP of ~425 mm) with elephant densities of 1.4 km⁻² (Asner & Levick

2012). The dominant tree in the elephant-modified savanna is *Colophospermum mopane*, an extremely vigorous resprouter. The scale of the elephant impacts is astonishing. For hundreds of square kilometres, the vegetation is reduced to a uniform shrubland of trees ~3 m tall on the gently undulating terrain with just a sparse scattering of taller individuals (Figure 8.7). Thickets of taller trees survive on small steep rock outcrops and steep river banks. It's hard to imagine that elephants were the culprits but the exclosure makes it clear that this is the case. Nevertheless, it is notable that elephant impacts are very heterogeneous even at the scale of Kruger

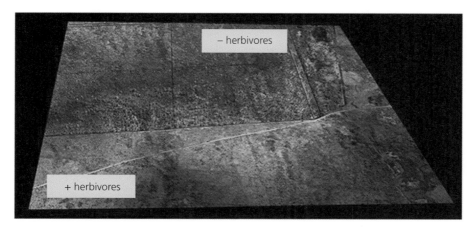

Figure 8.6 Airborne 3-D imaging of herbivore treatments in the Kruger National Park, South Africa. The treatments have been applied since the 1970s with a mesh fence enclosing low numbers of rare antelope but excluding all other herbivores larger than a hare. Tree breakage outside the exclosure indicates that elephant browsing is the main agent transforming the vegetation. Long-term basalt site (Nwashitsumbe). Colour-infrared spectroscopy highlights vegetation canopies (red) and dry/senescent vegetation and bare soil (blue to grey) overlain on the 3-D structure of each woody plant at a spatial resolution of 56 cm. From Asner et al. (2009). (See Plate 15 for colour version.)

National Park. On sandy granite soils in the more mesic south-west (MAP 675 mm), exclosure impacts were barely visible on the ground despite the same elephant densities (Figure 8.5; Asner & Levick 2012).

8.3.3 Natural experiments

Natural experiments are studies of situations where mammal herbivores have been added or removed or reduced from an area but not in a formal scientific context. Changes in plant communities in such instances are commonly attributed to changes in herbivore pressure. Other factors may also be involved and should be evaluated where data exists. The elimination of trees on small, remote islands is a much-repeated example of the devastating consequences for woody plants when goats, or deer, are introduced into virgin systems (e.g. Alfred Wallace's comments on St Helena).

A continental 'natural experiment' has been reported for Gorongoza, a protected area in Mozambique in southern Africa that was largely defaunated during a prolonged civil war. Gorongoza ecology was studied in the 1970s, with renewed research activity, after the war ended, since ~2015. Daskin et al. (2016) reported a large increase in trees, with the amount of increase varying in different habitats, since the

1970s. Floodplain environments showed the largest increase whereas miombo woodlands on nutrient-poor soils showed the least (Figure 8.8). They attribute the changes to the loss of browsers from the whole park, with no apparent directional trends in rainfall or fire over the period of tree increase.

The many studies of elephant impacts on savanna trees are not matched by elephant, or other herbivore effects, on forests and forest margins. The largest elephant impacts are typically in semi-arid savannas too dry for forest development. In Gabon, Jeffery et al. (2014) reported marked forest expansion into humid savannas in Lopé National Park. These observations are surprising because the park still had relatively high densities of the African forest elephant (*Loxodonta cyclotis*). The lack of significant elephant impacts on forest boundaries may also be true of mesic southern African forest/savanna mosaics (e.g. Boundja & Midgley 2010). Owen-Smith's (1987) hypothesis of keystone megaherbivore impacts on forest retreat, rather than on savanna tree cover, is yet to be tested in African savannas.

Beschta and Ripple (2009; Ripple & Beschta 2012) have reported extensively on the indirect effects of the return of wolves to Yellowstone. They were able to show that the age structure of deciduous trees

Figure 8.7 Mechanical damage to stems is more important than leaf loss for regulating tree populations. Top: heavily impacted *Colophospermum mopane* savannas in Kruger National Park. The trees' leader stems have been snapped off by elephants. Bottom: The same tree species forming woodlands when not impacted by elephants, Kruger National Park.

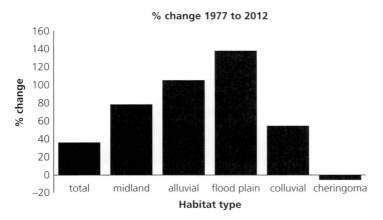

% change 1977 to 2012

Figure 8.8 Ecological legacies of defaunation during the civil war in Mozambique include large tree cover increases. Proportional tree cover change from 1977 to 2012 is shown for Gorongosa National Park. Change was not uniform but varied among habitats with the most heavily grazed and browsed habitats showing the largest tree increases. Cheringoma is a miombo woodland with poor quality forage where herbivores were historically rare. Redrawn from Daskin et al. (2016).

(*Populus* spp.; *Salix* spp.) reflected changing browse pressure coincident with the presence or absence of wolves. Return of wolves to the system in the 1990s has altered elk behaviour so that they have moved away from areas of intense browsing of aspen and willows. The result has been a large increase in tree sizes and densities of new recruits changing the physiognomy of the fertile valley habitat from open habitat to closed woodlands.

Introductions and re-introductions are a rich and under-utilized source of information on impacts of mammal herbivores or their predators. In South Africa, the (probably extra-limital) introduction of giraffe to Ithala Game Reserve caused extirpation of *Acacia davyi*, and near-extirpation of *Acacia caffra* except for a few trees taller than giraffe browse height, with *Acacia karroo* the next to go (Bond & Loffell 2001). However, some *Acacia* species were not visibly affected. These included *Acacia tortilis*, a very well defended tree that is common in heavily browsed habitats elsewhere in Africa. There were only 200 giraffe at Ithala in ~30 000 hectares (~0.6 giraffe km^{-2}). Much of the landscape is hilly and giraffe avoid steep slopes (they topple over). On gentler slopes, *Acacia davyi* and *A. caffra* were decimated by giraffe but the trees persisted on steep slopes. The lesson is that, even at low densities, megaherbivores can eliminate browse-sensitive species and therefore have a profound effect on tree species composition and tree densities. However, habitat heterogeneity (steep slopes) can preserve browse-sensitive species. Also, as for *Picea*, *Acacia drepanolobium*, or *A. tortilis*, not all trees are at risk.

8.4 Why herbivores impact plant populations

Herbivory has been intensively studied for many years. Models of herbivory were among the first to be developed in ecology. Nevertheless, models addressing the key question of whether herbivores can maintain open ecosystems, or open up closed ones, and under what circumstances, are in their infancy. There is a strong analogy to fire, a major generalist consumer. Both herbivores and fire regulate the probability of reaching and growing beyond size thresholds to reach full tree height. The browse

trap is analogous to the fire trap in that juvenile plants must grow above 'escape height' to grow the tree population. The impact of browsers varies with plant height. The browse height of antelope, deer, and caprids seldom exceeds 3m (du Toit 1990), which is similar to flame height in grass-fuelled fires. Below this height, plant growth rates can be impacted by browsers. Above the threshold height, browsers cannot reach leader stems and the sapling can grow into a tree. This is very similar to the problem of escaping the fire trap, the flame zone, in frequently burnt vegetation. The frequency of fires equates to the frequency of browsing which is determined by browser density. Fires vary in their intensity with more intense fires reducing height growth more than low intensity burns. Similarly browsers vary in their impact on the plant through the amount of stem they remove. Leaf loss has little impact on growth relative to pruning of shoots which alters the architecture of the plant. If leading shoots are pruned heavily, regrowth is often in the horizontal rather than vertical plane, delaying growth to adult size classes for years. Models generally assume that browser impacts are linearly proportional to leaf removal. Consequently they underestimate browse impacts on tree populations. Even small browser populations can radically alter size, structure, and height growth by snapping off leading shoots (Figure 8.7).

Palmer and colleagues, in a series of studies of red deer impacts on Scots pine recruitment in Scotland, provide insights into sensitivity of tree population growth to browsing (Chapter 3, Figure 3.6; Palmer & Truscott 2003). Escaping the browse trap is similar to escaping the fire trap but has not yet been modelled in an analogous way (Higgins et al. 2000; Hoffmann et al. 2012a). Growth rates of the pine in the absence of any browsing vary with habitat, taking 17 years to grow to 2m, the threshold height for escaping browsing of leader shoots, in this study. In poor conditions, saplings will take 32 years to reach escape height. If a plant is browsed, the probability of the leader shoots being eaten varies with height. As for fire, small saplings have a very high probability of losing the leader shoot if browsed, decreasing with height growth until they escape any browsing above ~1.5 m (Figure 3.6 a). The implication is that even light browsing at low deer densities will greatly diminish the probability of saplings escaping the browse trap to grow into trees. Under poor growing conditions, even low sporadic deer browsing will likely prevent any trees from maturing in this system. Our understanding of tree population regulation by consumers would benefit greatly from formal comparisons of how plants escape the fire trap versus the browse trap using the kind of information collected by Palmer and colleagues (Staver & Bond 2014).

8.4.1 Factors influencing stem loss

Plants can reduce stem loss to browsers by creating cage-like architectures through dense ramification, producing short shoots, and by developing stem spines which reduce feeding rates of many mammal browsers (Charles-Dominique et al. 2017b). Stem spines, in conjunction with cage architecture, greatly reduce stem and leaf losses during a browse event (Chapter 4). Losses also depend on the browser species and vary with its bite size and maximum stem diameter that can be bitten off. In one study, maximum stem diameters bitten off were proportional to body size of the browser, varying from 3.4 mm for bushbuck (20kg), a small-mouthed browser, to 6.5 mm for kudu (200 kg) and 19.4 mm for black rhino (900kg) (Wilson & Kerley 2003). Consequently, black rhino would have much larger architectural impacts on sapling escape rates than mesobrowsers even if the latter removed far more leaves in a browse event.

Elephants, or at least savanna elephants, not only strip leaves off trees but also snap off sizeable branches, especially the vertical leader shoots. In doing so, they alter the architecture of the tree, remove apical meristems, disrupt apical dominance, and change the developmental processes of the tree. The potential importance of stem loss for regrowth of browsed trees can be illustrated by analyses of elephant impacts on tree populations (Figure 8.5; Asner & Levick 2012; Asner et al. 2016). Stevens et al. (2016), in a study of woody encroachment in African savannas, found a general increase of trees except for conservation areas where elephants had been stocked for at least 20 years, where tree

populations were stable (Chapter 9). East Tsavo in Kenya is notorious for the near elimination of trees by elephants (e.g. Whyte et al. 2003).

Models of elephant impacts have been developed to indicate target densities above which elephant impacts might seriously reduce the tree layer. To illustrate the potential consequences of stem breakage on tree population growth, Figure 8.9 first shows an example of tree growth using different functions for how herbivory alters transitions to larger height classes in a four-stage matrix model. The linear response is that usually used where feeding impacts on plants is proportional to biomass consumed. The non-linear response assumes stem breakage slows transitions to larger size classes in non-linear fashion with increasing browser density (Figure 8.9; Moncrieff et al. 2014a). The model using the architectural assumption shows that elephants

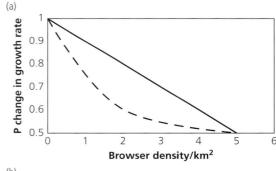

(a)

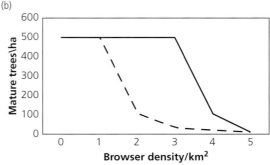

(b)

Figure 8.9 The effects of large herbivores on plant growth can be (a) linearly proportional to food intake and therefore to browser density (solid line), or non-linear where herbivores remove stems with buds changing the architecture of the plant (dashed line). (b) shows the simulated consequences for tree populations with the linear assumption (solid) showing impacts at double the browser densities of the non-linear assumption (dashed) in this example.

have much larger impacts on tree populations than the usual assumption of effects that are linearly proportional to biomass consumed. The point is that we lack models designed explicitly to explore the browse trap, rather than the fire trap. The processes involved are fundamental to understanding brown world dynamics and transitions from brown to green world.

8.4.2 Mechanical damage

If the consequences of removal of different plant parts is poorly understood, so too is the effect of mechanical damage caused by herbivores. For many years, herbivory research has emphasized food intake of browsers as the mechanism driving ecological impacts. Mechanical damage has only recently begun to be appreciated. Among the first to emphasize mechanical damage as a major ecological factor was Alan Savory in his seminal book on resource management of rangelands (Savory & Butterfield 1998). The use of intense trampling is a central feature of his livestock management system. Nevertheless, there are still very few studies of the ecological impacts of mechanical damage. Mechanical damage by elephants is a notorious cause of increased tree mortality in African savannas. It is not unusual for researchers from 'extinct' lands of the world to respond to their first exposure to an African megafaunal landscape with muted horror. Instead of a Garden of Eden, it may resemble a military training ground full of 'wanton' destruction, broken trees, smashed branches, and trampled shrubs. Lidar-based estimates of trees pushed over at Kruger National Park biennially are 12 per cent ha^{-1} (Asner et al. 2016)!

If trees are too large to be pushed over, they can still be killed by elephants stripping the bark, especially if the exposed inner layer is scorched by subsequent fires (Moncrieff et al. 2008). Mechanical damage is by no means characteristic of all elephant species. The violence to trees is mostly done by bull elephants and then mostly by the savanna species (*Loxodonta africana*). Females of this species strip leaves and break stems but seldom topple trees or snap off large branches (Midgley et al. 2005). In contrast, the forest elephants (*Loxodonta*

cyclotis) in Gabon cause negligible mechanical damage to trees and their browse impacts are not obvious (Terborgh et al. 2016a; Figure 8.10). African forest elephants may have more in common with the Asian elephant (*Elephas maximus*) in causing only minor mechanical damage to trees while feeding (Sukumar 1990; Terborgh et al. 2018). Both species also seem to lack the tree-toppling behaviour of African bull elephants. These behavioural differences among species suggest caution in using savanna elephants as models for the ecological impact of the extinct megaherbivores of the Americas, Europe, Australia, and Asia. They may be reasonable models. But we don't know.

Mechanical effects are by no means restricted to megaherbivores. Large herds of smaller browsers, even animals as small as springbok (*Antidorcas marsupialis*, 40 kg), can tear shrubs to pieces with their sharp hooves. Springbok herds of tens of thousands of animals migrated through arid shrublands in South Africa less than a century ago (Skinner 1993). The ecological consequences of such enormous herds trampling shrubs and grasses on their treks are largely unexplored. Significant trampling effects have been reported in Arctic shrublands during reindeer migrations, converting dwarf heath and mossy ground layers to grasslands (Zimov et al. 1995). Since natural migrations are threatened with extinction, this mode of herbivore impact is in serious need of urgent study to better understand indirect impacts of consumers.

8.4.3 Simulation models: putting herbivores into dynamic vegetation models (DVMs)

Simulation modelling offers another means of exploring impacts of herbivores on vegetation structure and determining key factors regulating their interaction. There are substantial difficulties in developing credible models of mammal herbivory. Among these is the problem of validation since, unlike fire, it is difficult to obtain a GIS layer of herbivore density or impacts against which to test simulations. Nevertheless, the first studies exploring herbivore impacts using DVM approaches have appeared (Pachzelt et al. 2015) and produced credible results on grazer abundance.

Figure 8.10 Top: Forest elephant (*Loxodonta cyclotis*) have relatively minor impacts on trees in Gabon forests. The arrow on the left points to two saplings that have been snapped off by browsing. The arrow on the right points to a stem that resprouted after a browse event. Bottom: Savanna elephants (*Loxodonta africana*) cause major damage to large trees such as this marula (*Sclerocarya caffra*).

8.5 Heterogeneity in herbivore impacts

8.5.1 Which herbivores?

Despite the impressive damage to tall trees caused by elephants, there is little evidence that they can convert a closed woody ecosystem into an open one. One possible example is from East Tsavo in Kenya where *Commiphora* woodlands with sparse and patchy grass were rolled back during a period of high elephant numbers (Whyte et al. 2003). *Commiphora* trees are brittle and generally fail to sprout when adults are knocked over. Their demise opened the system to a savanna supporting a different suite of large mammals preferring open habitats. Entirely different tree species replaced the *Commiphora*, dominated by spiny *Acacia* and *Terminalia* species. The Tsavo model of major elephant impacts on trees does not apply to the succulent thicket vegetation of South Africa. The dense thickets of Addo National Park retained their closed woody structure, if bisected by many paths, despite 50 years of elephant browsing at very high densities (~4 km^{-2}) (Lombard et al. 2001). Though woody plant composition was altered, the thicket structure persisted and was not replaced by open savanna except in very close proximity to watering holes. Interestingly, goat browsing outside Addo had much more extreme effects on the dominant shrub species than elephants in the park, apparently because of different feeding behaviour (Moolman & Cowling 1994). Goats are not native to this system. These contrasting examples indicate the heterogeneity of elephant impacts on vegetation. Quite different responses to the same browser species can occur depending on the attributes of the trees and the abiotic setting.

Structural alteration by forest elephants in Africa and India has not received the same level of study as savanna elephants. African forest elephants (*Loxodonta cyclotis*) survive in closed forests causing minor structural change relative to savanna elephants (*L. africana*). Though regular users of forest–savanna margins, forest elephants have failed to stop forests from advancing into the savannas in Gabon (Jeffery et al. 2014). So the kind of elephant, as well as the attributes of dominant trees, may also be 'context-dependent' when considering the role of megaherbivores in constructing open ecosystems.

8.5.2 Which plants?

Plant responses to herbivory, as illustrated by the Addo versus East Tsavo examples, are complex and there is no single, or not even a few, generalized plants for modelling or understanding plant response to herbivory. As is the case for fire, some plants are highly resistant to herbivory and these would be expected to predominate where herbivore pressure is high. Which plants survive in herbivore-dominated ecosystems depends on chemistry, structural defences, architecture, trampling and its consequences, and, for herbaceous plants, their ability to propagate under heavy grazing pressure (Box 8.1; Hempson et al. 2015a).

Box 8.1 Grazing lawns

The state of grasslands is an excellent indicator of the dominant consumers in grassy systems. Short-grass systems, or grazing lawns in tropical grassy ecosystems, indicate heavy grazing pressure (McNaughton 1984; Archibald et al. 2005). Contrary to the common perception in rangeland science that heavy grazing promotes woody plant increase, grazing lawns have very low woody cover. Woody plant establishment only occurs on lawns if grazers are excluded or move away to other pastures for extended periods. In tropical and subtropical grazing lawns, the grasses can spread despite being heavily grazed. They do so by spreading vegetatively by stolons, rhizomes, or flowering and setting seed with inflorescences bent close to the ground surface (e.g. Hempson et al. 2015b). Though highly productive in the wet season, grazing lawns epitomize 'brown world' in the dry season when the grasses are dried up. In the dormant season, they offer little or no forage and grazers are forced to utilize tall grass reserves left over from the wet season.

Grazing lawns generally have an associated fauna of short-grass feeders. In Africa these include the white rhinoceros (Figure 1; the black rhinoceros is a browser), the hippopotamus, blue wildebeest, impala (a mixed feeder), and warthog.

continued

Box 8.1 *Continued*

Exclusion experiments have shown that white rhinos are a key species for maintaining grazing lawns in productive mesic grassy systems but that other short-grass species maintain lawns in drier regions (Waldram et al. 2008). Hippos, like the rhinos, can also create lawns in highly productive grassy systems facilitating other short-grass feeders (Verweij et al. 2006). Exclosure experiments in Zululand lawns have shown that impala, which are mixed feeders, are the key species inhibiting woody plant recruitment. Mixed feeders build large herds feeding on grass in the wet season but switch to browsing as grass quality declines in the dry season.

Because short-grass systems offer only a seasonal grazing resource, they only form a fraction of grazing resources in any given landscape. For example, in Hluhluwe-iMfolozi Park (HiP), where white rhinos have been near carrying capacity for 50 years, lawns reached a maximum cover estimated as only 40 per cent of the area (Cromsigt et al. 2017). Thus 'brown world' is spatially heterogeneous with elements of black or green world contained within it. Grazing lawns are important, however, in preventing the spread of fire. Brown world, epitomized by short-grass swards, prevents the spread of fire and thereby conversion to black world. Only in a particularly wet year, or after successive years of high rainfall, will grass productivity outgrow the grazers and produce sufficient fuel to burn. Because of the suppression

of fire, and of woody plant recruitment, mammal consumers can create a distinct biome state with its own biota and recognizable from satellite imagery as a distinct state in Africa (Dantas et al. 2016).

Loss of lawns is a signal of major ecological change. Bond et al. (2001) identified the loss of lawns in HiP by noting a switch in the dominant acacias from *A. nilotica*, a well defended species, to *A. karroo*, a poorly defended species but one that is far more fire-tolerant than *A. nilotica*. The switch in species dominance followed a period of intense culling due to perceived overgrazing. This led to accumulation of grass biomass in the dry season fuelling large fires. The fires, in turn, eliminated *A. nilotica* but favoured the establishment of *A. karroo* seedlings. Thus, the system has switched from brown to black world with attendant changes in the biota. Grazing lawns have since contracted even more. Frequent fires draw animals away from lawns to the post-burn greening up swards reducing the grazing pressure needed to maintain short-grass swards. Lawns have contracted in the park to just a fraction of their previous extent and short-grass specialist species have followed them, reducing the apparent diversity of antelope species elsewhere. Fires in the park have not been successful in preventing forest encroachment. Thus, the system is in flux with shifts from brown to black to green world taking place and attendant species shifts.

8.6 Examples of brown world

Examples of OEs maintained by vertebrate herbivory were probably far more common in the Pleistocene when closed forest area was far smaller than today (Prentice & Harrison 2009). Flannery (1994) first made the case for the shrinking of 'brown world' and expansion of 'black world' after humans arrived in Australia and the megafauna went extinct. Contemporary Africa is the closest we have to a Pleistocene world before the mammal extinctions. However, each continent differed in climate, soils, vegetation, and also in their megafauna in the Pleistocene, with striking differences in body-size distribution (Svenning et al. 2016; Owen-Smith 2013). Reconstructing the legacy of these lost mammals is a fascinating task (Janzen & Martin 1982; Barlow 2000; Bond et al. 2004; Johnson 2006) and attracting growing research interest in the context of rewilding (Gill 2014; Bakker et al. 2016).

The problem with identifying brown world remnants in the modern world is that, unlike fires, they don't leave clear signatures in satellite imagery. Consequently, we lack a global perspective on the extent and characteristics of herbivore-controlled ecosystems. Hempson et al. (2015a) perhaps came closest to doing so by assembling a map of the 'herbivomes' in Africa. Herbivomes, analogous to pyromes (Chapter 7), are areas sharing similar functional groups of mammals. Five herbivore functional types were recognized, some quite well matched to particular biomes. For example, the small non-social browser functional type prefer closed forests while medium-sized, social mixed feeders (gazelles, for example) are characteristic of arid shrubby savannas in the south-west and north-east of Africa. The five functional types were grouped into four herbivomes. One of these, with the greatest abundance and functional diversity of mammals, is strikingly coincident with centres of distribution of spiny plants in Africa

(Charles-Dominique et al. 2016). These are the famed savanna parks such as the Serengeti and Kruger National Park. In contrast, a separate herbivome was identified distributed on deeply weathered land surfaces with dystrophic soils. These support mesic broadleaved miombo savanna woodlands where fire rather than grazers is the key consumer (Hempson et al. 2015a; Archibald & Hempson 2016). These high rainfall savannas are highly productive of grasses which have a low quality. During the long dry season, grass quality declines even further. Thus high rainfall C_4 grassy systems are the domain of fire, not herbivores.

Elsewhere in the world, areas of high herbivore impact are collections of case studies: similar to the situation with fire ecology before satellite mapping (e.g. Frank et al. 1998). The C_3 grasslands and shrublands of the cooler northern temperate regions are likely candidates for herbivore control over extensive areas. However, the interaction between fire and mammal herbivory has been less studied than in savannas. C_3 grasses remain green and edible through the lean season with cold rather than drought limiting growth. The steppes of Eurasia, including Mongolia, may be examples of consumer control since they support, or did support, large herds of grazers until recently. Consumer control of steppe grasslands in Eurasia, especially in the vast forest/steppe mosaics (Erdős et al. 2018) requires more study. Global fire maps show an arc of high fire activity in these C_3 grasslands (Chapter 7), but it is not clear how heavy grazing, fire, and climate interact in driving the vegetation of this vast region.

Montane systems in the northern hemisphere support extensive alpine pastures maintained by grazing. The world's largest pastoral alpine ecosystem is the *Kobresia* ecosystem on the Qinghai Tibetan plateau (Miehe et al. 2019). The system covers an area of 450 000 km², and supports 5 million people, 13 million yaks, and 30 million sheep and goats. In addition there are a number of endemic large mammals, including wild yak (*Bos mutus*), chiru (an antelope, *Pantholops hodgsonii*) and kiang (*Equus kiang*). In addition are numerous pika (*Ochotona*, a lagomorph), a grazer similar in size and lifestyle to North American prairie dogs. In these high elevation sites, treelessness might be attributed to cold and climate extremes. However, numerous, if widely scattered, groves of trees indicate that most of the pastures are below the tree line (Figure 8.11). Palynological studies indicate that extensive forests occurred in the region in the early Holocene, switching to grasslands

Figure 8.11 Montane uncertain ecosystems. *Juniperus przewalskii* groves surrounded by grazer-maintained *Kobresia* pastures ('alpine meadow') in the Tibetan highlands, the world's largest pastoral alpine ecosystem (Qinghai Province, 3800 m; Miehe et al. 2019). Photo G. Miehe (1998).

in the mid-Holocene when the climate was milder! The causes of the change are debated but fire is implicated. Today the grasslands and sedgelands (*Kobresia* is a sedge) are grazed too short to burn by the numerous grazing herds. Unlike savannas, the system seems to function without having reserves of taller grasses to maintain grazers through the lean season. Similar grazer-maintained montane pastures are common in Eurasia (Miehe et al. 2019).

In western Europe, cage-like architectures of open-habitat woody plants have been observed in European meadows and interpreted as anachronisms persisting after extinction of the megafauna (Bakker et al. 2004, 2016). They survive today where domestic livestock have access to the meadows. Cattle and horses are domesticated descendants of the wild aurochs and tarpan respectively. If livestock are excluded, there is gradual succession to broad-leaved forest, an example of a switch from brown to green worlds. *Quercus* species recruit in the shelter of the spiny shrubby clumps. These oaks are light-demanding and cannot recruit in the shade of closed, shady forests. Vera (2000; Vera et al. 2006) has argued that these open habitats with bush clumps, 'wood pastures', were common in Europe before megafaunal extinction. There is some evidence that fires also maintained open habitats in Western Europe. The hypothesis of a more open pristine Europe is controversial and has stimulated numerous studies exploring the openness of the landscapes in the past (e.g. Svenning 2002; Mitchell 2005). It has also stimulated experimental studies of the impacts of large grazers on woodlands and meadows and challenged management objectives for conservation lands (Vera 2009).

Australia is dominated by highly flammable ecosystems and is quintessentially 'black world' (Orians & Milewski 2007). Flannery (1994) first proposed major shifts from herbivore-dominated vegetation to fire-dominated vegetation as a result of megafaunal extinction after the arrival of humans in Australia. His hypothesis has stimulated much subsequent work elsewhere on competing consumers. Australia's megafauna went extinct some 45 ka, at least 30 ka earlier than the Americas and Eurasia. Thus there has been more time for fire systems to replace mammal-dominated systems. Surviving native herbivores have also suffered from the introduction of alien predators, starting with the dingo. Unlike the mainland, Tasmania, until recently, had no introduced predators. It is also the only part of Australia where grazing lawns have been reported (Leonard et al. 2010). These are maintained by wombats and wallabies. *Acacia* species near the lawns have cage-like architectures quite different from chemically defended acacias that are common on the mainland. Wallaby browse lines are evident on several tree species near the lawns (Hazeldine & Kirkpatrick 2015). Tasmania also has green world elements mingled with black world flammable shrublands and woodlands (Jackson 1968; Bowman & Wood 2009; Wood & Bowman 2012). Is this a model for pre-human Australia (cf Flannery 1994), or the oddities of island life?

8.7 Alternative states and regime shifts

8.7.1 Herbivory and fire as competing consumers

Animal activity at a site can be assessed by direct counts or indirectly in various ways of which dung counts have proved particularly useful. Fire activity is now widely quantified by satellite imagery but fires have been mapped for decades before that and changes in closed forest boundaries have been mapped using aerial photography since the 1930s. Figure 8.12 shows the abundance of dung plotted against number of fires for transects in a protected

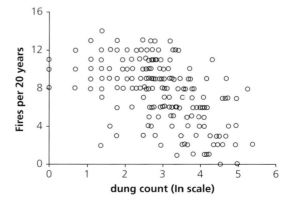

Figure 8.12 Black versus brown world in an African savanna park (Hluhluwe Imfolozi). Fires were recorded as the number of fires in 20 years while herbivore pressure was recorded by dung counts. Each point represents a sampling transect. Data from Staver et al. (2012).

area in a South African savanna. The data show either high herbivory and low fire activity or high fire activity and relatively low herbivore activity (Staver et al. 2012). Vegetation is consumed primarily by fire or by herbivory though neither consumer completely excluded the other. In a sequence of wet years, with high grass productivity, there is sufficient grass to carry fires in brown world sites. In frequently burnt tall grass areas, grazers also consume grass but not enough to stop fires spreading. In this acacia-dominated savanna, there are clear distinctions in plant architecture between brown and black worlds. Species in heavily grazed areas have cage-like architectures well protected from browsers but vulnerable to fire damage. In frequently burning black world habitats, acacias have tall pole-like architectures enabling rapid growth above flame height (Chapter 4; Archibald & Bond 2003; Staver et al. 2012). The divergent architectures are most prominent in saplings. Tree species above 'escape' height (~2–4m for both fire and ungulates) converge in their architecture whereas their saplings might differ profoundly. Regime shifts can be identified by changes in juvenile versus adult tree species (Bond et al. 2001). Just such shifts have been reported in this system with disparate species composition of adults and saplings reflecting a landscape-scale shift from brown to black worlds.

Regime shifts can also reflect political change. With the break-up of the Soviet Union in the early 1990s, there was a steep decline in livestock grazing (mostly sheep) in a 19 000 km^2 area of steppe grasslands in southern European Russia (Dubinin et al. 2011). Livestock numbers dropped from ~900 000 in 1985 to less than 200 000 in 1999. The reduction in livestock numbers lead to increased grass biomass, fuelling fires. From 1985 to 1996, burnt area was very low with a maximum of 20 000 ha in one year. From the mid 1990s, fires started burning larger areas reaching a maximum of 320 000 ha a decade later, 16 times larger than during the Soviet era. Similarly large changes in burnt area have been reported in Valencia, Spain, following rural depopulation and resultant changes in fuel continuity as farmers flocked to cities (Chapter 9; Pausas & Fernández-Muñoz 2012). Regime shifts between black and brown world can occur rapidly and over diverse scales.

Continental scale:

Archibald and Hempson (2016) were the first to explore consumer domains at the continental scale in sub-Saharan Africa (Figure 8.13; Plate 16). They obtained fire data from satellite imagery and herbivore densities from regional data on livestock density fitted to environmental parameters to generate 'herbivore surfaces'. Consumer domains were obtained from the raw data on percentage burned and kg herbivores. Green world pixels were those where both burnt area and herbivore biomass were below the median value, brown world where herbivore biomass but not burnt area was greater

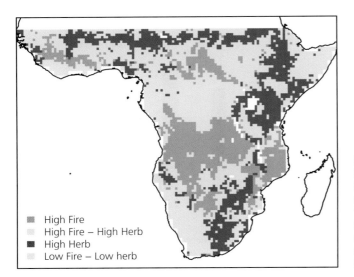

High Fire
High Fire – High Herb
High Herb
Low Fire – Low herb

Figure 8.13 Consumer realms for Africa. High fire = black world, high herbivory = brown world, low fire—low herbivory = green world. Mixtures of black and brown world at the mapping scale are also shown. Note that 'green world', at equilibrium with climate, includes tropical forest but also arid shrublands in the south-west and north-east that are too dry to support closed woody ecosystems. From Archibald and Hempson (2016). See Plate 16 for colour version.

than the median value, and black world where burnt area but not herbivore biomass was greater than the median value. Pixels where both burnt area and herbivore biomass were above the median were classified as black and brown. Figure 8.13 shows the results divided into black, brown, and green world across Africa.

Archibald and Hempson (2016) also compared biomass consumed by fires with biomass consumed by herbivores adjusted for their body mass. The dominant consumer in open ecosystems varied along a rainfall gradient with fire predominating in more humid regions and herbivores at the more arid end. Figure 8.14 shows changing patterns of consumption by fire or vertebrate herbivory in different rainfall zones. The pattern indicates bistability: most pixels have either fire or herbivory as the dominant consumer with few intermediates. Consequently, areas of mixed consumer dominance in Figure 8.14 are chequerboards of brown and black world with the patches too small to map at the continental spatial scale (but see Figure 8.13). Soil nutrients were a secondary influence on consumer domain with fires predominating on poorer soils and herbivores on richer ones.

8.7.2 Stability of states and regime shifts: paleoecological evidence

There is now clear evidence for the existence of different states of open ecosystems maintained by different classes of consumers at diverse scales (Archibald & Hempson 2016; Dantas et al. 2016). Both fire and herbivory maintain open ecosystem states. Furthermore, new studies are showing that each state has distinct suites of species and that these have distinct functional attributes, depending on the dominant consumer (Chapter 4; Staver et al. 2012; Charles-Dominique et al. 2015b). But ASS theory also predicts that each state is persistent (stable) yet also able to switch to alternative states. Evidence for the stability of brown world systems has emerged from *Sporormiella*, a coprophilous fungus with distinctive spores, used as a proxy for herbivore density. Calibration studies have shown that *Sporormiella* is a good correlate of modern mammal densities within ~100 m radius of the collection site. In sediment cores, shifts from high counts of *Sporormiella* to high counts of charcoal indicate regime shifts from brown to black world. They have been observed in sites as divergent as North America, Madagascar, and Australia (Robinson et al. 2005; Burney et al. 2003; Gill et al. 2009; Rule et al. 2012). In Australia, for example, a remarkable core from Lynch Creek in Eastern Australia spans 130 ka into the past (Figure 8.15). High *Sporormiella* counts early in the sequence declined sharply at 41 ka, a few thousand years after humans first settled in Australia (~45 ka). Human settlement was followed by megafaunal extinction during a period when climates were not changing. The reduction in mammal herbivory is marked by a steep decline in *Sporormiella*. Grass

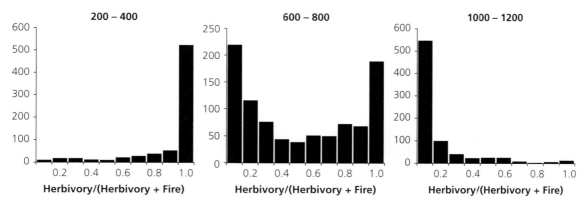

Figure 8.14 Fire and herbivores are 'competing consumers' in Africa. Proportional consumption by herbivores in three rainfall bands (200–400, 600–800, 1000–1200 mm MAP) showing the tendency for NPP in a pixel to be consumed either by herbivores or by fire. From Archibald & Hempson (2016).

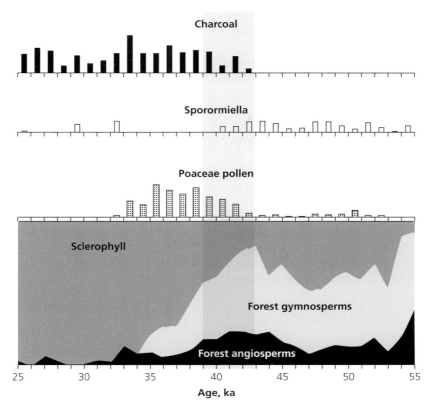

Figure 8.15 Paleo-evidence for stability, and regime shifts, of brown to black world, Lynch's Crater, Australia. *Sporormiella* is a coprophilous fungal spore and a proxy for mammal herbivores. After humans settled in Australia, many large mammal species became extinct from ~41 to 39 ka. Grass pollen increased, fires increased as shown by charcoal. As fires continued, fire-maintained sclerophyll vegetation came to dominate the pollen signal and grass pollen declined. Closed forests occurred in the landscape with gymnosperm pollen dominating when herbivory was high and fires rare. After the regime shift from herbivores to fire, with increasing dominance of sclerophylls, forest gymnosperm pollen disappeared. Redrawn from data in Rule et al. (2012).

pollen then increased, followed by a large increase in charcoal indicating a surge of fire activity. The pollen record indicates shifts from predominantly open habitat with a significant fraction of 'green world' forest conifers (*Podocarpus* spp.) before the extinctions, followed by flammable grasslands which were then replaced by flammable sclerophyll woodlands characteristic of much of Australia today. Thus, the Lynch Creek record indicates a three-coloured world with stable dominance of brown world for thousands of years, apparently along with *Podocarpus* closed forest, followed by a regime shift to black world when the vertebrate consumers collapsed. Paleoecological evidence of this kind for stability of alternative states is rare. North American

examples are complicated by climate changes at the end of the Pleistocene as the system also changed from abundant herbivores to frequent fire and then to a closed forest. Nevertheless, paleoecological studies can provide critical evidence for the stability of vegetation states, their potential to switch to alternative states, and changes in consumers associated with these shifts.

8.8 Summary and conclusion

This chapter has emphasized how herbivores can regulate tree populations helping to prevent woody plants from recruiting to form dense populations. Though herbivory has been studied for many years, the browse trap and its implications for herbivore regulation of tree populations is still understudied. It is clear that there is enormous variability in herbivore impacts. The source of this variability is in browser properties such as feeding height but also in plant properties and the nature of defences against different feeding modes of animals. The abiotic template also causes considerable heterogeneity in browser impact, for example in creating plenty of refugia in hilly country but none in flat terrain. Perhaps the most striking knowledge gap is the near absence of studies on animal impacts on boundaries between open and closed ecosystems (but see Bakker et al. 2004; Vera et al. 2006). Such studies could be very informative on the circumstances favouring forest expansion or contraction under sustained browsing or grazing. The ecological effects of really big animals, the megaherbivores, both now and in the past, are frustratingly difficult to identify. Unlike fire, which leaves at least some trace of its impacts on plants in the fossil record, the mammal impact is hidden. Even the largest herbivores of all time, the long-necked (sauropod) dinosaurs, have left frustratingly little trace of how they ate plants. As Bakker (1978) suggested 40 years ago, dinosaurs could have been enormously influential in altering the structure of vegetation and the evolutionary trajectory of plants. But the ecological interactions of ancient giant vertebrate herbivores (tetrapods) and ancient plants is largely a closed book. This is in striking contrast to the exponential growth of knowledge on ancient fires and their ecological and evolutionary impacts on plants (Belcher 2013). There remains much to be done in the realm of vertebrate herbivory as moulder of ecosystems.

CHAPTER 9

The future of open ecosystems

What is the future of open ecosystems? Here I am not referring to secondary grasslands and scrub in abandoned old lands or successional stages towards secondary forests. What is the fate of the ancient savannas, grasslands, shrublands, and woodlands that are the central topic of this book? Since, by definition, they are far from equilibrium with climate, these ecosystems have the potential for massive transformation. For the same reason, standard approaches for projecting the future under climate change are doomed to failure. Unsurprisingly, the biggest threat for untransformed open ecosystems is their conversion to forests, whether by invasion of native trees, alien invasive trees, or afforestation schemes. In this chapter, I consider the distinctive threats to the future of open ecosystems and some of their consequences for the goods and services they provide.

I first consider the consequences of rising atmospheric CO_2, the key indicator of human impacts on the Earth system. While most concern is rightly on the impacts of CO_2 on climate, there are direct CO_2 effects on plants that will upset the balance between trees and shrubs, threatening the future of OEs. Trees are increasing in previously open grassy systems worldwide, a process called woody encroachment. Though elevated CO_2 is likely a significant contributor to bush encroachment, so are changes to consumers such as suppression of fires, or too much or too little herbivory. Fire and herbivore management has been studied for decades, but the Anthropocene brings new challenges in managing trajectories of change. Climate change, and direct effects of CO_2 increase, will have very different effects on grassy versus woody OEs. Maintaining fires in grassy systems will be increasingly difficult as trees increase and the landscape is fragmented by roads, croplands,

and settlements, inhibiting fire spread. In contrast, fires in woody OEs are more likely to thrive in the future, burning under ever more extreme weather conditions. Large budgets and expensive technology has failed to suppress these 'megafires' which cause major damage to people and properties. The megafires have increased sharply in the new millennium to frequencies unprecedented in the historical record. In the writings of ancient Greece, the classical texts hardly mention fires. What has changed?

Ecological considerations for the future of OEs apply to untransformed lands. But the biggest threats in the Anthropocene will surely come from pressure to transform wildlands into fields for crops. New agricultural methods are opening up possibilities for industrial-scale agriculture on the leached soils of the tropics, threatening the future of natural ecosystems. A growing threat to OEs is also their targeting for afforestation to sequester carbon, a short-term response that puts off serious action on global warming for a few more years. But will planting trees have the desired effect of cooling the planet? The science is surprisingly thin and the processes sufficiently complex to challenge the best of science communicators explaining the nuances to the public.

9.1 The end of savannas?

9.1.1 CO_2 effects

Savannas were 'born' in low CO_2 atmospheres in the Oligocene. For the last 30 million years, plants have evolved under low CO_2. Fossil fuel burning changed CO_2 concentrations from 285 pp in 1850 to nearly 300 ppm by 1900, rocketing up in the last century to 410 ppm by 2018. CO_2 is projected to reach 700 ppm

Open Ecosystems: ecology and evolution beyond the forest edge. William J. Bond, Oxford University Press (2019). © William J. Bond 2019.
DOI: 10.1093/oso/ 9780198812456.001.0001

or more by the end of the century based on current emissions (IPCC). Such high CO_2 concentrations were last seen in the Eocene, more than 35 million years ago, when broadleaved forests covered most of the world (Beerling & Royer 2011). Will savannas disappear at such high CO_2? From modelling studies, Collatz et al. (1998) predicted the future of C_4 versus C_3 plants. According to their study, C_4 grassy systems will have shrunk from 60 per cent of the world's land surface at the last glacial maximum (LGM, CO_2 at ~200 ppm) to < 10 per cent by the end of the century (CO_2 > 700 ppm). Higgins and Scheiter (2012), using a DGVM tuned to African ecosystems, predicted that the climates occupied by savannas would shrink from their maximum extent along a precipitation gradient in the LGM to survive only in the arid extremes by the end of the century (Figure 9.1) because of CO_2 effects on trees versus grass. In contrast, Sage and Kubien (2003) analysed the future for C_4 grasses, noted their declining performance under high CO_2, but considered that fires would increase to such an extent in a globally warming world that C_4 grassy systems might expand! Contrary to climate-only projections (Bergengren et al. 2011), these large uncertainties reflect the complexity of interacting controls in open ecosystems and how to model them.

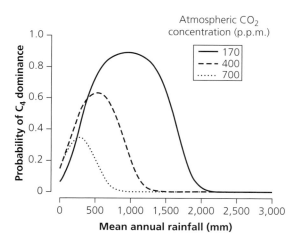

Figure 9.1 The probability that a site will be C_4 dominated (C_4 grassland or savanna) as a function of rainfall for different atmospheric CO_2 concentrations. The data are from DGVM simulations for Africa. The three CO_2 concentrations represent the last glacial maximum 20 ka BP (170 ppm), 2015 (400 ppm), and 2100 (700 ppm). From Higgins & Scheiter (2012).

Many studies cite differences in photosynthetic pathway as a factor underlying CO_2 effects on woody thickening (C_3 trees versus C_4 grasses). However, I know of no study of tree–grass coexistence that has considered differences in photosynthetic rate as the mechanism underlying competitive performance. It is what plants do with the carbon that counts: woody versus herbaceous plants, regardless of photosynthetic pathway, behave very differently. The relative performance of C_3 versus C_4 plants under changing temperature and CO_2 is best compared ecologically between plants with the same growth form—grasses with different photosynthetic pathways.

9.1.2 Woody encroachment

Increases in woody plants, especially in the last half-century, have been reported worldwide for grassy biomes (Archer et al. 2017; Stevens et al. 2017). The evidence comes from satellite imagery, aerial photographs, long-term monitoring of plots, and photo comparisons of vegetation change over time. Rates of change vary, often increasing with precipitation so that mesic areas are at greatest risks of tree invasion. The causes of woody encroachment are complex (e.g. O'Connor et al. 2014; Archer et al. 2017; Devine et al. 2017). They include fire suppression, grazing (too heavy or too light), no browsers, replacement of native megafauna with domestic livestock, reduced human use of wood for fuel and building, and climate variability. There is growing evidence that increasing atmospheric CO_2 is an additional contributor. Polley (Polley et al. 1997) was among the first to recognize that increasing CO_2 could lead to an increase in tree recruitment. His proposed mechanism was reduced transpiration because of reduced stomatal conductance under elevated CO_2. Reduced water use by plants should result in increased soil moisture and longer growth periods. Indeed increased soil moisture is one of the most common results reported in elevated CO_2 experiments in grasslands (Morgan et al. 2004). The effect of increased soil moisture should be deeper infiltration which would benefit deep-rooted woody saplings more than grasses (Polley et al. 1997).

Increasing CO_2 also increases photosynthetic rate if there is a sink for the extra carbon fixed. If the surplus carbon cannot be used, photosynthesis is 'down-regulated' and there is no CO_2 effect on carbon gain. Common carbon sinks in an ecological context are starch reserves in below-ground storage organs, clonal expansion using carbon allocated to root or stem extension, woody stem growth, and rapid resprouting after injury (Figure 9.2). Faster

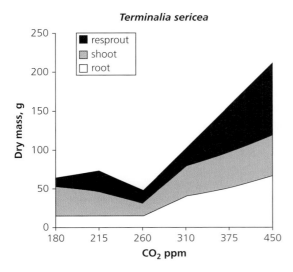

Figure 9.2 The response of *Terminalia sericea*, an African savanna tree, to a gradient in CO_2. Plants were grown for a year, cut to simulate fire (shoot) and harvested after a second growing season (resprout, root). From Bond & G. Midgley (2012).

sapling growth and greater resprouting ability should feed into increased tree recruitment in grassy systems, more frequent sapling escape from the fire or browse trap, better drought survival, and therefore woody thickening. At low CO_2, characteristic of the last glacial period, DGVM simulations showed that sapling growth rates would be so slow that no sapling would survive fire, browsing, or drought to grow into established trees, and savannas would be replaced by grasslands (Bond et al. 2003b). The simulated growth advantages for South African savanna tree species were close to observed growth rates for plants grown in a glasshouse (Kgope et al. 2010). Similarly strong responses to CO_2 and drought survival were observed in *Sequioa* seedlings exposed to low CO_2 (Quirk et al. 2013). As yet, no field manipulation of CO_2 has been done, whether in open-topped chambers or FACE experiments. However, if increasing CO_2 does increase seedling survival and sapling release from consumer control, then these responses should be visible in long-term fire experiments. Buitenwerf et al. (2012) analysed responses to a 50+ year fire experiment in Kruger National Park. Figure 9.3 shows per cent changes in different height classes from the 1970s censuses after 20+ years of the fire treatments compared to the 2003 census after another ~30 years of fire treatments. In this mesic savanna, the dominant tree, *Terminalia sericea*,

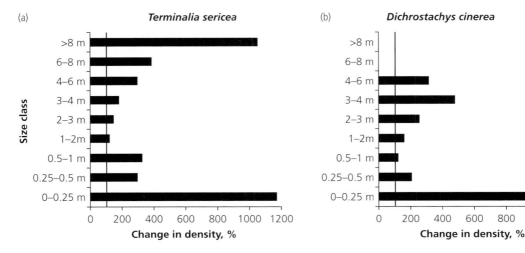

Figure 9.3 Savanna plants subjected to the same fire regime for >50 years have increased in density. The increase for *T. sericea* is consistent with glasshouse measures of CO_2 responses (Figure 9.2). (a) shows % change in density across size classes for the dominant savanna tree and (b) the same for the dominant shrub. Data is from identical areas sampled 30 years apart and subjected to the same fire treatments since the 1950s in experimental burn plots, Kruger National Park, South Africa. Adapted from Buitenwerf et al. (2012).

showed a ten-fold increase in seedlings and an eight-fold increase in trees that had escaped the fire trap (>3m). The dominant shrub, *Dichrostachys cinerea*, a clonally spreading species, also showed very large increases across size classes. These changes are consistent with glasshouse studies showing strong CO_2 responses of *Terminalia sericea* to a CO_2 gradient. Similar strong responses were found in *Acacia karroo*, a highly invasive native species in South African savannas (Buitenwerf et al. 2012; Bond & G. Midgley 2012). These studies strongly suggest that increasing CO_2 is an additional factor, along with fire, herbivory, climate variability, and human wood harvesting causing widespread bush encroachment in African savannas.

CO_2 effects have also been reported in other savannas. Hoffmann et al. (2000) were the first to test CO_2 responses in a savanna tree using a species from the Brazilian cerrado. Plants were grown at 350 and 700 ppm for up to 25 weeks. The seedlings responded positively to elevated CO_2 but, when clipped and fertilized, showed very strong biomass increases. If the interaction with fertilizer turns out to be more general for savanna plants on dystrophic soils, then ecological impacts of CO_2 might be stronger in species from drier and/or nutrient-rich savannas but muted on heavily leached nutrient-poor soils.

These studies have focussed on savannas. However there is also widespread evidence for expansion of forest patches in savanna/forest mosaics. Forest expansion is occurring despite frequent grass-fuelled fires in diverse areas, including Gabon (Jeffery et al. 2014), Northern Australia (Bowman et al. 2010), and South Africa (Wigley et al. 2010, Beckett 2018). Mechanisms of spread have not been explored beyond Hoffmann et al.'s (2012a) model of tree thresholds at the forest margins with savannas. Given their large root systems, burnt trees on the forest margin should have a very large sink for carbon and would benefit greatly from elevated CO_2 allowing rapid recovery. It would be intriguing to expose a burnt forest margin to elevated CO_2 to observe whether post-burn recovery is accelerated. But beyond the scientific fascination is concern about savannas being swallowed up by expanding forests (Jeffery et al. 2014). There is also concern as to whether forest will retreat because of more frequent fire storms linked to more frequent extreme weather as the world warms (Beckett 2018). Global change impacts at the forest edge warrant more study.

9.2 The future of consumers that maintain open ecosystems

Humans have already eradicated megaherbivores first in Australia, then in the Americas, followed by Eurasia, and now focussing on Africa. There is no question that megaherbivores are awkward, inconvenient, and usually dangerous neighbours and, understandably, most communities would prefer to keep them behind large fences. Fragmentation of their habitat is inevitable as human populations grow and urbanize. Fire is intrinsically anarchic to settled, ordered, human communities and enormous resources have been used to suppress this last functional 'megaherbivore'. Fire in grassy biomes may be rendered harmless or suppressed altogether. But in woody fuels, the trend is for larger, ever more dangerous fires and, potentially, the expansion of open ecosystems into pyrophobic forests. The trajectories of change, and possible management options, are so different for fire in grassy versus woody fuels that they are best treated separately.

9.2.1 The end of savannas? Fire management

Grass-fuelled fires are generally of rather low intensity and easily contained by firebreaks, roads, croplands, and settlements. People can control the intensity by choosing to burn early in the dry season when grasses are not fully cured. Even the time of day matters, with less intense burns in the afternoon and evening. This ability to manage fires can be observed at continental scale in comparisons of densely settled and farmed savanna landscapes such as West Africa versus the sparsely settled savannas of North Australia (Archibald et al. 2013). In the former, fires are small, early season, and of low intensity. In Australia, with similar fuels, fires are large, late season, and of much higher intensity. These are the two pyromes for grassy fuels identified by Archibald et al. (2013) (Chapter 7). Fire suppression of grassy fuels was commonly practised in the USA in the twentieth century leading to large-scale conversion of pine savannas (Noss 2012; Gilliam

& Platt 1999) and oak savannas to closed forests (Nowacki & Abrams 2008).

A similar biome switch, but to tropical forests, is currently happening in Brazil. Cerrado remnants in an increasingly cultivated landscape seldom burn and the result has been transformation to closed ecosystems in the last few decades (Durigan & Ratter 2016; Pinheiro & Durigan 2009). The effort needed to burn the remnants so as to retain the rich diversity of the cerrados is enormous given the number and small size of most of the remaining fragments. Furthermore, until recently, national legislation promoted the suppression of fire and still does so to some extent (Durigan & Ratter 2016). Similar examples of fragmentation of flammable open ecosystems are legion in the more densely settled tropics. In the Anthropocene, at least in grassy ecosystems, fire may suffer the same fate as the Pleistocene megafauna, surviving only in protected areas that retain a culture of burning.

The problem of reduced fire is exacerbated where tree growth has increased because of CO_2 fertilization. Fire regimes that effectively controlled trees in the past are far less effective today (Figure 9.3). Rangeland ecologists have begun exploring the use of high-intensity fires to reclaim grass in bush-encroached areas. In parts of the USA (Texas, Kansas, Oklahoma), ranchers have formed conservancies to re-introduce fire as a cost-effective means of restoring the productivity of their rangelands. Here neighbours celebrate when a fire encroaches from a neighbouring property instead of going to the courts to litigate (Twidwell et al. 2013; Kreuter et al. 2008). Under extreme weather conditions—high temperatures, low humidity, high windspeed—and given sufficient dry grass fuels, high-intensity fires can spread from grassy ecosystems and cross biome boundaries penetrating deep into forests. In South Africa, conservation managers attempting to reclaim savannas which have been invaded by scrub forests are exploring similar methods for creating high-intensity fires during high-risk fire weather conditions (Smit et al. 2016; Beckett 2018; Figure 9.4). Areas subjected to these high-intensity fires have first to be sealed off by ordinary prescribed fires around their margins to contain them. Follow-up burns are required to promote grasses and prevent forest trees from resprouting and recovering to shade out the

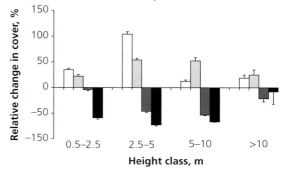

Figure 9.4 High intensity fires may help reduce bush encroachment. This example, from Kruger National Park, compared changes in woody plant cover with successively increasing fire intensity (white, control; light grey, low-; dark grey, moderate-; black, high-intensity). High intensity fires were most effective at reducing cover of smaller woody plants (<2.5m). Redrawn from Smit et al. (2016), *J. Appl. Ecol.*, 53, 1623.

grasses again. These extreme measures may become more common out of necessity as the ecological balance shifts from grass to trees in a high CO_2 world.

9.2.2 The end of savannas? The future of the megafauna

In an analysis of bush encroachment in north-eastern South Africa, the only sample areas which showed stable, or declining, tree cover from the twentieth to the twenty-first century were conservation areas with elephants (Figure 9.5). Mesic savannas, those with climates suitable for forests, closed up with trees regardless of the presence of elephants (Stevens et al. 2016). At least for the semi-arid savannas of Africa, this study suggests a more wooded future as the disastrous decline of elephants as a result of the illegal ivory trade continues across the continent. The massive poaching pressure on rhinos, particularly on the docile white rhinoceros, a mega-grazer, also has major ecological repercussions. The white rhino is an ecosystem engineer helping to maintain grazing lawns in mesic savannas (Waldram et al. 2008; Cromsigt & te Beest 2014). Extirpation of the rhino has been shown to result in conversion of short fire-resistant grazing lawns to tall flammable bunch grasslands. Fire has proved less effective than herbivory in controlling bush encroachment so that

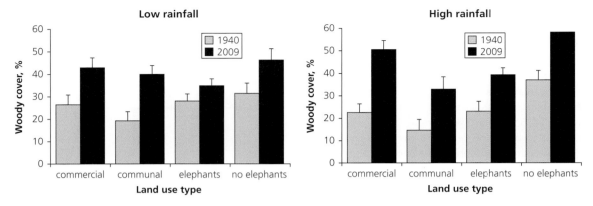

Figure 9.5 Woody encroachment in savannas in relation to land use in low (<650 mm MAP) and high (>650 mm MAP) rainfall regions of South Africa. All areas showed a significant increase in woody cover from 1940 to 2009 except low rainfall areas with elephants which showed no significant increase. Areas with no elephants are conservation areas stocked with indigenous fauna. From Stevens et al. (2016).

the extirpation of the rhino and other short-grass grazers will indirectly lead to a shift to more wooded savannas and then closed thickets (scrub forest).

In African savanna systems, increasing tree cover has cascading consequences. Reduced visibility increases predation of the megafauna (Le Roux et al. 2018). Fear of predation in low visibility bush areas alters use of habitat by plains grazers such as zebra and wildebeest (Valeix et al. 2009). In Kruger National Park changes in tree densities have been correlated with changing composition of herbivores, with declines in open habitat species vulnerable to predation and increases in species that can tolerate predation risk as woody plants increase (Smit & Prins 2015). Thus, management of the last remnants of Africa's magnificent megafauna is increasingly characterized by the problem of invasive native woody plants. These systems will require more intense management to sustain the full complement of large mammal species.

Birds are also sensitive to habitat structure. Open habitat birds are threatened by increasing tree cover and marked reduction of the ranges of some species (e.g. secretarybird, *Sagittarius serpentarius*, and southern ground hornbill, *Bucorvus leadbeateri*) have been observed in South Africa over the last 20 years as woody cover in savannas continues to increase (Sirami et al. 2009; Hofmeyr et al. 2014). Unlike warming effects in temperate and boreal climates, the biota of open ecosystems in EUCZ are threatened

by changes in vegetation structure, a poorly appreciated and poorly researched consequence of global change.

9.2.3 The expansion of open woody ecosystems? Megafires and their control

'Not all fires can be stopped.... treating wildfires like other disasters such as earthquakes has value. No one considers stopping earthquakes... rather communities develop infrastructure to make living in earthquake country safer.' **J. Keeley, 2012**

Large fires depend on fuels that are sufficiently continuous for fires to spread. In grassy systems, continuous fuel beds depend on sufficient rainfall to drive high grass productivity. Fires decline in drought years because of lack of grass fuel. In woody vegetation such as Mediterranean-type shrublands, fuel accumulation occurs over many years. Extreme fire weather creates continuity in fuels. In woody systems, fuel continuity is greatest after long droughts when the shrubs and other fuels dry out enough for fires to spread regardless of local microclimate and local variation in fuel properties. Major fires occur under hot, dry, and especially windy conditions similar to fire storms in savannas. However, in striking contrast to savannas, they usually occur after protracted drought causes fuels to dry out on a landscape scale. Such megafires have become common in many regions in the 2000s, including southern

Table 9.1 Big fire events in Mediterranean climate regions with woody fuels. Updated from Keeley et al. (2012) for northern California and Chile.

Region	Year	Size ha	Losses		Ignition sources	Wind	Weather, drought
			Buildings	**Lives**			
Portugal	2003	430 000	200	21	Lightning	High wind	Heat wave
Greece	2007	271 000	2100	84	Arson, accident	High winds	Heat wave
California							
northern	2017	14 900	5643	22	Power lines	High winds	Severe drought
southern	2007	109 500	2400	15	Power lines	High winds	Severe drought
Central Chile	2017	279 930				High winds	Severe drought
South Africa							
West Cape	2000	8000	8	0	Accidental	High winds	Alien plant fuels
Australia							
W. Australia	1961	40 000	160	0	Lightning	Cyclonic winds	
Victoria	2009	330 000	2029	173	Power lines	High winds	Heat wave, drought

Europe, the western USA, eastern Australia, Chile, India, and other regions with woody fuels (Williams 2013; San-Miguel-Ayanz et al. 2013). These high-intensity fires cause major damage to people and property (Table 9.1).

Enormous resources are used in attempts to contain them. In the USA, for example, the US Forest Service spends more than half of its budget on fire-fighting at the expense of funds for forest management. In 2017, fire suppression costs exceeded $2 billion in attempts to control fires in the west, Pacific Northwest, and Northern Rockies (USDA Press Release No. 0112.17). Climate change projections indicate that these megafires are likely to become more common in future (Westerling et al. 2011; Flannigan et al. 2013). Thus, in striking contrast to savannas, open wooded ecosystems may spread into closed woody ecosystems if the frequency of extreme weather conditions increases.

9.2.4 Large fires and land-use change

Countries bordering the Mediterranean Sea raise interesting questions about competing consumers in woody, rather than grassy, vegetation. Elephants occurred in Europe until a few thousand years ago, and persisted on islands until ~4000 BP. Hannibal used elephants in his attack on Rome ~2200 years ago, though whether they were native to North

Africa is unlikely. Did the megafauna in the past, including the megaherbivores, open up the dense shrublands of the Mediterranean? Did animal herds reduce fire activity by breaking up the continuity of fuels? With domestication of cattle, sheep, goats, equids, the extinct megafauna would have been rapidly replaced. Classical literature from the region seldom mentions fires as an environmental risk (Papanastasis et al. 2010). Papanastasis et al. (2010) suggest that fires were rare because of high densities of browsers, especially goats. Heavy browsing by livestock still continues but has declined in Europe because of changing land use, while still prominent in North Africa. In Europe, rural depopulation has emptied the countryside of small-scale farmers along with their herds. Pausas and Fernández-Muñoz (2012) explored the consequences of rural depopulation for changing fire regimes. They compiled a 150-year record of fires in Valencia in southern Spain. Their study showed an abrupt increase in fire activity from the 1970s (Figure 9.6). They attribute these changes to depopulation, leading to shrub colonization of old lands and pastures. In addition, foresters established plantations of flammable pines. The switch from fuel-limited to weather-limited constraints on fire spread suggests Mediterranean shrublands may, like savannas, have had competing consumers producing mosaics of black and brown worlds (Chapter 8). Comparisons of the fire regime in Valencia with

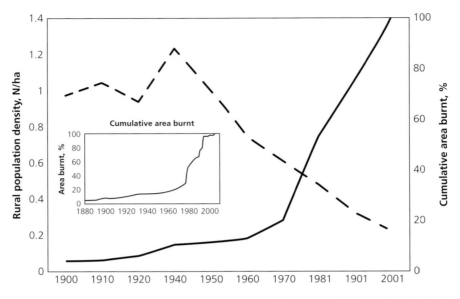

Figure 9.6 Rural depopulation in Valencia, Spain was followed by large increases in burnt area. Dashed line, rural population densities, individuals/ha; solid line, cumulative area burned, %. The inset shows the cumulative area burned dating back to 1880. The switch in fire regime began in the 1970s and is attributed to increased fuel continuity following land abandonment, livestock reductions, and afforestation with pine trees. Redrawn from data in Pausas & Fernández-Muñoz (2012).

Morocco, where traditional land use has continued (until recently), supported the arguments of Pausas and Fernández-Muñoz (2012) since fire activity was very low in both regions, diverging only in the past few decades as land abandonment accelerated in Spain (Chergui et al. 2018). Recent suggestions that rewilding may help contain the huge fires, where advanced and expensive technology has failed, thus has some empirical support (Johnson et al. 2018)!

9.3 Invasives and open ecosystems

The commonly held idea that organisms are optimally adapted to their environment, implicit in DGVMs, has been undermined by new comparative studies of biomes (Lehmann et al. 2014), the characteristics of their major growth forms in different regions (Moncrieff et al. 2014c), and the climates under which they occur (Moncrieff et al. 2016). It has also been undermined by observations on the spread of alien invasive plants. For example, the idea that plants growing on very nutrient-poor soils are uniquely adapted to such soils cannot be sustained when these same soils are invaded by alien

grasses and woody species that lack such adaptations. Similarly, the idea that low and erratic rainfall has selected for succulent growth forms as the optimal solution to the climate cannot be sustained when the Sonoran succulent desert vegetation is converted to highly flammable grassland by an invasive African grass, *Cenchrus ciliaris*.

Open ecosystems are particularly vulnerable to major structural changes because they are far from the climate-limited tree 'carrying capacity' of closed forests. Furthermore, the dominant native growth forms are typically intolerant of shade and very vulnerable to invasive trees. In the Cape region of South Africa, fynbos, the major vegetation type in this, the world's richest temperate flora, occurs on soils extremely poor in nutrients. Many fynbos species have highly specialized nutritional strategies designed to 'mine' the very low quantities of phosphorus in the soil (Lambers et al. 2008). Yet despite the poor soils, fynbos is readily invaded by a variety of trees including *Pinus* spp., Australian acacias, *Hakea* spp. (Proteaceae), and Australian Myrtaceae (Richardson & Rejmánek 2011). The acacias are highly productive, fix nitrogen, access a deeper rooting volume, and accumulate phosphorus into the surface soils (Yelenik et al. 2004). Native plants are shaded out by the acacias and are very difficult to restore to the nutrient-enriched sites where they have been cleared. Pine trees, escaped from plantations, are common invasive trees on steep mountain slopes where they eliminate most of the native species. Pines have ectotrophic mycorrhiza and, despite the very nutrient-poor soils, form dense stands.

Since 1995, South Africa has developed a major programme of controlling invasive species, especially trees. The programme employs tens of thousands of people, as a form of job creation, costs nearly one billion rand per annum, and goes by the name 'Working for Water' (Turpie et al. 2008). It is not the biodiversity impacts of the invasive trees but their impact on water, a critical resource for the country, that motivates this major public works programme. Hydrological research in the mountain catchments of South Africa had revealed that afforestation of open ecosystems, shrublands in the winter rainfall areas, C_4 grasslands in the summer rainfall regions, reduces streamflow. The research design was to select paired catchments, monitor streamflow for a period of years, afforest one of the pairs, and measure any changes in streamflow relative to the control. The results were unequivocal; afforestation reduced streamflow, drying out the rivers (Bosch & Hewlett 1982). These results remain a major motivation for maintaining open ecosystems in mountain catchment areas. They led to legislation restricting afforestation where the reduction in streamflow would negatively impact established downstream communities. Subsequent studies in many parts of the world have confirmed these findings. Generally, forests use more water and yield less to streams and rivers, than open, non-forested ecosystems (Jackson et al. 2005). Thus, one serious consequence of the widely observed encroachment of forests into grasslands and open shrublands is reduced water supply.

9.4 Habitat transformation

Untransformed open ecosystems face many threats under the multiple impacts of anthropogenic forcing. However, by far the greatest threat is likely to be their transformation to croplands, forestry plantations, or human settlements. In Brazil, cerrado often occurs on deep, highly weathered red soils (oxisols). Such soils are widespread in the humid tropics. Unlike the post-glacial landscapes of North America, Eurasia, China, and Argentina, these soils are intrinsically infertile. They have long resisted conversion to commercial-scale agriculture. However Brazilian and other agricultural scientists have developed new crop genotypes, fertilizer regimes, and agronomic practices to create, for the first time, industrial-scale agriculture on these soils. The result has been a major economic boost to Brazil, revitalizing depopulated rural communities but also causing unprecedented transformation of the cerrado into croplands (Lapola et al. 2014). Brazil is actively promoting the extension of the system to similar climates and soils elsewhere in the tropics, starting with Africa. The scale of transformation is vast (Figure 9.7). It is almost entirely restricted to savannas, open ecosystems, and deforestation of tropical forests has declined steeply in recent years (Lapola et al. 2014). Conservation biologists in Brazil are beginning to grapple with legislation for conserving 'wildlands' and how to manage the process to preserve some kind of functioning remnants of these remarkable savannas (Durigan & Ratter 2016). The lessons learnt will be of great value to other regions of the tropics where large-scale transformation is technically feasible.

9.5 Carbon sequestration, afforestation, and reforestation

9.5.1 Global tree-planting programmes

Open ecosystems are natural targets for afforestation because woody biomass is below climate potential. Plantation forestry has flourished in the southern hemisphere by planting up OEs in high rainfall grasslands and shrublands. Large forestry industries were built up in Australia, South Africa, Chile, and New Zealand, usually based on *Pinus* species, a genus which does not occur in the southern hemisphere. Conflicts over water and with alternative land uses have slowed the spread of forestry plantations. But a new impetus for afforestation has developed in attempts to reduce global warming by planting trees to sequester carbon. Carbon sequestration projects have become a major enterprise in the

Figure 9.7 An example of the widespread transformation of cerrado, the world's most biodiverse savanna, to crop monocultures in Brazil. Infertile tropical red soils were avoided for crop farming until recently. Agricultural breakthroughs on several fronts have opened up these lands to industrial-scale cropping. The innovations are spreading to similar climate and soil conditions elsewhere, including higher rainfall savannas of Africa. Photo credit: Pedro Biondi/ABr. License: Creative commons attribution 3.0 br.

tropics marketed as favouring the environment by reversing the tide of deforestation. International programmes, such as the Bonn challenge, have been supported by the IUCN, governments, and the World Bank. The Bonn challenge aims to 'reforest' 1.5 million km^2 by 2020 and 3.5 million km^2 by 2030. The 2030 target is an area larger than France, Germany, Italy, UK, Greece, Sweden, Norway, Denmark, and the Netherlands combined, or about half the area of the Australian continent!

A significant problem, though seldom discussed, is deciding where to plant the trees. If open ecosystems are of anthropogenic origin, then all open ecosystems are potential targets for tree planting. If, however, open ecosystems include ancient systems, with their own unique biota, then afforestation is as brutal a transformation as industrial agriculture. The World Resources Institute has mapped areas of the world which they believe to be suitable for afforestation on the massive scale targeted by the Bonn challenge (the Atlas of Forest Restoration Opportunities; https://www.wri.org/resources/maps/atlas-forest-and-landscape-restoration-opportunities). The map is based on comparing actual versus climate potential woody biomass. Not too surprisingly, the Atlas picks out the major OEs of the world: African grassy biomes, South American cerrado, the forest/steppe mosaic of Eurasia, and the grassy biomes of Indo-Malaysia. As implied by its name, the Atlas does not distinguish between reforestation and afforestation. The Atlas makers ignored any evidence for antiquity of OEs or ancient consumer-control. For example, the map of 'degradation' and 'deforestation' uses a definition of 'degradation' as any process that damages trees and reduces their biomass or cover. This elevates trees above all other life forms. Consequently, iconic savanna parks of Africa, the Serengeti and Kruger National Park, are mapped as 'deforested' or 'degraded'. Elephants and fire damage trees so, by definition, consumer control by these agents represents degradation. The Atlas has not been revised, or withdrawn, despite critical comment on its failure to distinguish between reforestation and afforestation (Veldman et al. 2015a, b; Bond 2016). How can you 'reforest' areas that have not been forest for millions of years? The Atlas continues to form the basis of massive tree planting programmes.

Real deforestation is widespread in the tropics and has accelerated in some regions in recent decades (e.g. Hansen et al. 2013). The problem for environmentally sensitive afforestation projects is to distinguish secondary grasslands produced by deforestation from ancient, primary, open ecosystems. Veldman et al. (2015a) pointed to markers of 'old growth' grasslands which can be used to identify these systems and to help focus on their conservation values. These include high species richness, especially of forbs, high endemism to open habitats, and long-lived perennial plants, often with large underground storage organs that resprout rapidly after frequent fires (see Chapter 4). The list of indicators is preliminary because the study of many of these systems is still in its infancy in most parts of the world. Consequently, it is hard to generalize from the few detailed studies that have been made. Furthermore, in comparison to tropical forests, very few researchers are working on tropical open ecosystems (Bond & Parr 2010; Parr et al. 2014).

The old growth concept emphasizes conservation values of open ecosystems. However, OEs are also of great value as catchment cover reducing erosion and yielding more water than plantation forests (Bosch & Hewlett 1982; Jackson et al. 2005; Honda & Durigan 2016). Open ecosystems—grasslands, shrublands, and savannas—included the grazing lands of the Pleistocene megafauna. Domestic livestock have replaced much of the extinct (and, in Africa extant) megafauna in ecosystems that are often well adapted to herbivory. Especially in drier regions, these are the natural ecosystems supporting livestock production. Ironically, the crops produced by large-scale conversion of Brazilian cerrado are largely used for fattening livestock in feedlots of the industrial world (soya bean, maize, African fodder grasses) or providing vegetarian food substitutes (soy milk, soy flour, soy protein, tofu etc.).

9.5.2 Will afforestation of open ecosystems cool the world?

Planting trees not only stores carbon above ground—it also changes biophysical properties of the vegetation and most notably the albedo (the fraction of sunlight reflected back into space). Snow, for example, has a very high albedo, reflecting most of the sunlight received back into space. Dark surfaces, such as the canopy of boreal conifers, absorb sunlight. Consequently, tree planting in tundra would warm, not cool, the Earth (Betts 2000; Betts et al. 2007). C_4 grasslands also have highly reflective surfaces, especially in the dry season and compared to forests and plantations. Converting grasslands to forests would be expected to have a warming effect, opposing the net cooling effect of any carbon accumulated. Grasslands store carbon below ground where it is more resilient to drought, fire, and rising temperatures than trees. Recent studies in California suggest that grasslands are more reliable carbon sinks than forests because of the risk of losing trees to wildfires and drought (Dass et al. 2018). The net effect of major landcover changes, from open to closed ecosystems, needs far more study (Bonan 2016). Feedbacks to the climate from landcover changes include differences in evapotranspiration of open versus closed forests, aerosol particles from smoke, and the spatial extent of transformation which is critical for estimating the magnitude of any climate feedbacks (Beerling & Osborne 2006; Bonan 2008; Zhao & Jackson 2014). There is generally a remarkable imbalance between the vast amount of work on carbon sequestration and the scarcity of studies on biophysical feedbacks of landcover change (Bright et al. 2015). Given the uncertainties, it seems premature to use climate change as a motivation for major landscape transformation of open ecosystems to forests. This is especially the case when old-growth OEs are converted to plantations using carbon sequestration as motivation. Attempts at restoring tropical old-growth grasslands are at least as difficult, if not more so, than restoring forests in deforested areas (Box 9.1). If the Bonn challenge gets anywhere near its target of 'reforesting' 3.5 million km² by 2030, it will ruin the biodiversity of old-growth grasslands for centuries, and will likely

---

Box 9.1 Restoration of open ecosystems

Restoration of forests after deforestation has been practised for many years and methods are well advanced (Chazdon 2017). Restoration of grassy systems is more recent and has proved difficult because of the distinctive life histories of their constituent species. Passive restoration of old-growth grasslands failed to restore the rich forb diversity of a tropical grassland, even after 20 years of recovery time. Forbs (non-graminoid herbs) are long-lived perennials characterized by large underground storage organs (USOs). They are common in frequently burnt grasslands with wet summers. Forbs with USOs resprout very rapidly after fire and typically flower and fruit in the first year post-burn before being smothered by grasses. Seedbanks are rare or non-existent while seed set of the long-lived species is usually very poor. These forbs are very well adapted to frequent fires but not to habitat destruction such as clearing for crops or planting up to forest plantations. Passive restoration is ineffective for such plants because of the lack of seed and poor seedling survival. The most successful species are the few short-lived plant species with high seed production and dormant seedbanks. Planting of turfs has been attempted. Though slow and expensive, the transplants do result in reasonable survival but there is little spread from the turf to the grasslands in which it has been planted. Grasses that first colonize restoration sites often persist for decades without being replaced by later successional species that dominate untransformed grasslands. Given the current state of knowledge, restoration is not an option for recovering species-rich, old-growth grasslands. The alternative is to develop criteria for recognizing old-growth grasslands and then avoiding their transformation by mining, agriculture, or afforestation. The fragility of these rich diverse grasslands, unfortunately, implies that quick-fix solutions to global warming of extensive afforestation will leave a legacy of impoverished secondary grasslands for decades, if not centuries. For an excellent review of the current state of knowledge on restoration of grasslands, see Buisson et al. (2018).

---

trigger a wave of extinction to the light-loving biota of the world.

9.6 Conclusions

This book has been about the ecology and biogeography of systems that, unlike forests, do not even have a collective name. However they occupy all the vegetated continents and support large fractions of the world's population (Parr et al. 2014). Among the challenges facing researchers on OEs is how to recognize them. They occur, by definition, in climates that can also support forests. So when is an area too dry or too cold for forests to develop? It is no use trying to identify them with correlative methods, however accessible remote sensing and the associated analytical methods have become. An area may be 95 per cent grassland. But that does not make it a grassland climate. It is the 5 per cent that isn't grassland that indicates the climate potential for forest. Foresters have often made the experiment of planting trees in open ecosystems. The results are invaluable in showing the extent of climate potential to grow trees. Forestry also reveals the importance of phylogenetic constraints on tree growth, such as the remarkable eucalypts (Lehmann et al.

2014). The development of DGVMs is a very useful pointer to areas climatically suitable for trees but where they may be absent. It can be hard for the modellers to accept, but really bad fits of a model simulation to field data are also very revealing as to areas of ignorance that need further exploration.

Having identified an area as one where ecosystems are uncertain, one is set for exploring why it is closed or open. There is a great deal of research on the ecology of grasslands, savannas, shrublands, and forests. A constant problem in writing this book has been straying from the problem of explaining openness to the ecology of the open system. We know far less about the interface between closed and open biome states than we do about each state. People tend to work in one or the other system with little dialogue across the boundary. The same is true of trophic levels. We tend to work in just one or, at most, two trophic levels. The trophic perspective used in this book has been productive and insightful. Including fire, with large vertebrate herbivores, as major consumers opens up new windows on similarities and differences in how they shape vegetation. We are beginning to understand how plants and animals differ in functional traits, and the cascading consequences of adding or subtracting a key consumer. Exploring across systems, and across trophic levels, enriches ecological experience and insights while opening up new questions and new perspectives.

Paleoecology is interesting in its own right. But when you are exploring mosaic vegetation, potential alternative states, the discipline becomes essential. Paleoecology can reveal both stability at a site and the potential, and rate, of state change. Diverse proxies for fire, and large herbivores, have been developed and provide powerful evidence for long-term stability of a state, and of regime shifts to alternative states (even if it does take a lot longer than aquatic systems). There are still significant problems in exploring the scale of mosaic vegetation, especially using traditional cores: pollen and charcoal move at different rates, leading to uncertain inferences. Reconstructing vegetation mosaics is an active area of research. Back in deep time, ancient fires have been studied using diverse approaches with growing revelations on fire impacts on Earth history (Belcher 2013; Scott et al. 2014; Scott 2018). However, reconstructing large vertebrate impacts on ancient vegetation has lagged. If you are at all familiar with a modern patchwork of green, black, and brown world, then it is easy to imagine something similar in the Cretaceous. Angiosperm-fuelled fires would

have swept over some landscapes, ornithischian dinosaurs grazed down stubble elsewhere, and forests that escaped both consumers would persist in refugia. But what filled the role of grasses? How important is the appearance of new growth forms, new innovations, in the changing ecologies of the past? Imagine the world without grass. What occupied all those 'grassland climates' before the grasses existed?

A major motivation for writing this book was concern over ill-considered plans to afforest huge areas of OEs in the next decade as a quick fix for global climate change (Veldman et al. 2015b; Bond 2016). That quick fix could do more damage to the biota of OEs then centuries of human endeavour. Afforestation policies entirely disregard the science of OEs that has developed at a remarkable pace over the last 20 years. The irreversible damage, in our lifetimes, of planting up old-growth tropical grasslands to trees, for example, can only be justified after careful weighing up of the costs and benefits: what will be lost, and who will gain. Will we plant a tree and destroy an ancient grassland? Or plant a tree to restore a broken forest? There is a world of difference. After centuries of neglect, scientists are shaking off cultural baggage and opening their minds to the non-forested parts of the world. Maybe it is not too late.

References

Abis, B., & Brovkin, V. (2017). Environmental conditions for alternative tree-cover states in high latitudes. *Biogeosciences*, 14(3), 511–27.

Agarwal, I., & Ramakrishnan, U. (2017). A phylogeny of open-habitat lizards (Squamata: Lacertidae: *Ophisops*) supports the antiquity of Indian grassy biomes. *Journal of Biogeography*, 44(9), 2021–32.

Aleman, J. C., Jarzyna, M. A., & Staver, A. C. (2018). Forest extent and deforestation in tropical Africa since 1900. *Nature Ecology & Evolution*, 2(1), 26.

Alves, R. J., Silva, N. G., Junior, F., Aluisio, J., & Guimaraes, A. R. (2013). Longevity of the Brazilian underground tree *Jacaranda decurrens* Cham. *Anais da Academia Brasileira de Ciências*, 85(2), 671–8.

Andersen, A. N. (2018). Responses of ant communities to disturbance: five principles for understanding the disturbance dynamics of a globally dominant faunal group. *Journal of Animal Ecology*. https://doi.org/10.1111/1365-2656.12907

Andersen, A. N., Cook, G. D., & Williams, R. J. (2003). *Fire in tropical savannas: the Kapalga experiment* (Vol. 169). Springer Science & Business Media.

Andersen, A. N., Woinarski, J. C., & Parr, C. L. (2012). Savanna burning for biodiversity: fire management for faunal conservation in Australian tropical savannas. *Austral Ecology*, 37(6), 658–67.

Anderson, J.M., & Spencer, T. (1991). *Carbon, nutrient and water balances of tropical rain forest ecosystems subject to disturbance: management implications and research proposals.* MAB Digest 7, UNESCO, Paris.

Anderson, R. C. (2006). Evolution and origin of the Central Grassland of North America: climate, fire, and mammalian grazers. *Journal of the Torrey Botanical Society*, 133(4), 626–47.

Archer, S. R., Andersen, E. M., Predick, K. I., Schwinning, S., Steidl, R. J., & Woods, S. R. (2017). Woody plant encroachment: causes and consequences. In: D. D. Briske (ed.), *Rangeland systems: processes, management and challenges*, (pp. 25–84). Springer, Cham.

Archibald, S., & Bond, W. J. (2003). Growing tall vs growing wide: tree architecture and allometry of *Acacia karroo* in forest, savanna, and arid environments. *Oikos*, 102(1), 3–14.

Archibald, S., Bond, W. J., Stock, W. D., & Fairbanks, D. H. K. (2005). Shaping the landscape: fire–grazer interactions in an African savanna. *Ecological Applications*, 15(1), 96–109.

Archibald, S., & Hempson, G. P. (2016). Competing consumers: contrasting the patterns and impacts of fire and mammalian herbivory in Africa. *Philosophical Transactions of the Royal Society of London B: Biological Sciences*, 371(1703), 20150309.

Archibald, S., Lehmann, C. E., Belcher, C. M., Bond, W. J., Bradstock, R. A., Daniau, A. L.,…& Higgins, S. I. (2018). Biological and geophysical feedbacks with fire in the Earth system. *Environmental Research Letters*, 13(3), 033003.

Archibald, S., Lehmann, C. E., Gómez-Dans, J. L., & Bradstock, R. A. (2013). Defining pyromes and global syndromes of fire regimes. *Proceedings of the National Academy of Sciences of the United States of America*, 110(16), 6442–7.

Archibald, S., Roy, D. P., van Wilgen, B. W., & Scholes, R. J. (2009). What limits fire? An examination of drivers of burnt area in Southern Africa. *Global Change Biology*, 15(3), 613–30.

Archibald, S., Staver, A. C., & Levin, S. A. (2012). Evolution of human-driven fire regimes in Africa. *Proceedings of the National Academy of Sciences of the United States of America*, 109(3), 847–52.

Asner, G. P., & Levick, S. R. (2012). Landscape-scale effects of herbivores on treefall in African savannas. *Ecology Letters*, 15(11), 1211–17.

Asner, G. P., Levick, S. R., Kennedy-Bowdoin, T., Knapp, D. E., Emerson, R., Jacobson, J.,…& Martin, R. E. (2009). Large-scale impacts of herbivores on the structural diversity of African savannas. *Proceedings of the National Academy of Sciences of the United States of America*, 106(12), 4947–52.

Asner, G. P., Vaughn, N., Smit, I. P., & Levick, S. (2016). Ecosystem-scale effects of megafauna in African savannas. *Ecography*, 39(2), 240–52.

Augustine, D. J., & McNaughton, S. J. (1998). Ungulate effects on the functional species composition of plant communities: herbivore selectivity and plant tolerance. *Journal of Wildlife Management*, 62(4), 1165–83.

Bailey, R. G. (2014). *Ecoregions: The Ecosystem Geography of the Oceans and Continents*. Springer-Verlag, New York.

Bakker, E. S., Gill, J. L., Johnson, C. N., Vera, F. W., Sandom, C. J., Asner, G. P., & Svenning, J. C. (2016). Combining paleo-data and modern exclosure experiments to assess the impact of megafauna extinctions on woody vegetation. *Proceedings of the National Academy of Sciences of the United States of America*, 113(4), 847–55.

Bakker, E. S., Olff, H., Vandenberghe, C., De Maeyer, K., Smit, R., Gleichman, J. M., & Vera, F. W. M. (2004). Ecological anachronisms in the recruitment of temperate light-demanding tree species in wooded pastures. *Journal of Applied Ecology*, 41(3), 571–82.

Bakker, R. T. (1978). Dinosaur feeding behaviour and the origin of flowering plants. *Nature*, 274(5672), 661–3.

Balch, J. K., Brando, P. M., Nepstad, D. C., Coe, M. T., Silvério, D., Massad, T. J.,…& Cury, R. T. (2015). The susceptibility of southeastern Amazon forests to fire: insights from a large-scale burn experiment. *Bioscience*, 65(9), 893–905.

Barlow, C. (2000). *The ghosts of evolution: nonsensical fruit, missing partners, and other ecological anachronisms*. Basic Books, New York.

Barlow, J., & Peres, C. A. (2004). Ecological responses to El Niño-induced surface fires in central Brazilian Amazonia: management implications for flammable tropical forests. *Philosophical Transactions of the Royal Society of London B: Biological Sciences*, 359(1443), 367–80.

Barrett, P. M., & Willis, K. J. (2001). Did dinosaurs invent flowers? Dinosaur–angiosperm coevolution revisited. *Biological Reviews*, 76(3), 411–47.

Beadle, N. C. W. (1954). Soil phosphate and the delimitation of plant communities in eastern Australia. *Ecology*, 35(3), 370–5.

Beadle, N. C. W. (1966). Soil phosphate and its role in molding segments of the Australian flora and vegetation, with special reference to xeromorphy and sclerophylly. *Ecology*, 47(6), 992–1007.

Beckage, B., & Ellingwood, C. (2008). Fire feedbacks with vegetation and alternative stable states. *Complex Systems*, 18(1), 159.

Beckage, B., Platt, W. J., & Gross, L. J. (2009). Vegetation, fire, and feedbacks: a disturbance-mediated model of savannas. *The American Naturalist*, 174(6), 805–18.

Beckett, H. (2018). Firestorms in a mesic savanna-forest mosaic. PhD dissertation. University of Cape Town, South Africa.

Beerling, D. (2007). *The emerald planet: how plants changed Earth's history*. Oxford University Press, Oxford, UK.

Beerling, D. J., & Osborne, C. P. (2006). The origin of the savanna biome. *Global Change Biology*, 12(11), 2023–31.

Beerling, D. J., & Royer, D. L. (2011). Convergent cenozoic CO_2 history. *Nature Geoscience*, 4(7), 418.

Beerling, D. J., Taylor, L. L., Bradshaw, C. D., Lunt, D. J., Valdes, P. J., Banwart, S. A., Pagani, M., & Leake, J. R. (2012). Ecosystem CO_2 starvation and terrestrial silicate weathering: mechanisms and global-scale quantification during the late Miocene. *Journal of Ecology*, 100(1), 31–41.

Belcher, C. M. (ed.) (2013). *Fire phenomena and the Earth system: an interdisciplinary guide to fire science*. Wiley-Blackwell, Oxford, UK.

Belcher, C. M., Collinson, M. E., & Scott, A. C. (2013). A 450-Million-Year History of Fire. In C. M. Belcher (ed.) *Fire phenomena and the Earth system: an interdisciplinary guide to fire science*, 229–49. Wiley-Blackwell, Oxford, UK.

Belcher, C. M., & Hudspith, V. A. (2017). Changes to Cretaceous surface fire behaviour influenced the spread of the early angiosperms. *New Phytologist*, 213(3), 1521–32.

Belcher, C. M., Mander, L., Rein, G., Jervis, F. X., Haworth, M., Hesselbo, S. P.,…& McElwain, J. C. (2010). Increased fire activity at the Triassic/Jurassic boundary in Greenland due to climate-driven floral change. *Nature Geoscience*, 3(6), 426.

Belsky, A. J. (1990). Tree/grass ratios in East African savannas: a comparison of existing models. *Journal of Biogeography*, 17 (4–5), 483–9.

Bennett, B., & Kruger, F. (2016). *Forestry and Water Conservation in South Africa*. ANU Press, Canberra, Australia.

Bennett, K. D. (2009). Woodland decline in upland Scotland. *Plant Ecology & Diversity*, 2(1), 91–3.

Bergengren, J. C., Waliser, D. E., & Yung, Y. L. (2011). Ecological sensitivity: a biospheric view of climate change. *Climatic Change*, 107(3–4), 433–57.

Beschta, R. L., & Ripple, W. J. (2009). Large predators and trophic cascades in terrestrial ecosystems of the western United States. *Biological Conservation*, 142(11), 2401–14.

Betts, R. A. (2000). Offset of the potential carbon sink from boreal forestation by decreases in surface albedo. *Nature*, 408(6809), 187.

Betts, R. A., Falloon, P. D., Goldewijk, K. K., & Ramankutty, N. (2007). Biogeophysical effects of land use on climate: Model simulations of radiative forcing and large-scale temperature change. *Agricultural and Forest Meteorology*, 142(2–4), 216–33.

Binkley, D., & Valentine, D. (1991). Fifty-year biogeochemical effects of green ash, white pine, and Norway spruce in a replicated experiment. *Forest Ecology and Management*, 40(1–2), 13–25.

Binkley, D. A. N., & Giardina, C. (1998). Why do tree species affect soils? The warp and woof of tree-soil interactions. In van Breemen, Nico (Ed.) *Plant-induced soil changes: Processes and feedbacks*, (pp. 89–106). Springer, Dordrecht.

Bonan, G. B. (2008). Forests and climate change: forcings, feedbacks, and the climate benefits of forests. *Science*, 320(5882), 1444–9.

Bonan, G. B. (2016). Forests, climate, and public policy: A 500-year interdisciplinary odyssey. *Annual Review of Ecology, Evolution, and Systematics*, 47, 97–121.

Bond, W.J. (2005). Large parts of the world are brown or black: a different view on the 'Green World' hypothesis. *Journal of Vegetation Science*, 16, 261–6.

Bond, W. J. (2008). What limits trees in C4 grasslands and savannas?. *Annual Review of Ecology, Evolution, and Systematics*, 39, 641–59.

Bond, W. J. (2010). Do nutrient-poor soils inhibit development of forests? A nutrient stock analysis. *Plant and Soil*, 334(1–2), 47–60.

Bond, W. J. (2015). Fires in the Cenozoic: a late flowering of flammable ecosystems. *Frontiers in Plant Science*, 5, 749.

Bond, W. J. (2016). Ancient grasslands at risk. *Science*, 351(6269), 120–2.

Bond, W. J., Cook, G. D., & Williams, R. J. (2012). Which trees dominate in savannas? The escape hypothesis and eucalypts in northern Australia. *Austral Ecology*, 37(6), 678–85.

Bond, W. J., & Keeley, J. E. (2005). Fire as global 'herbivore': the ecology and evolution of flammable ecosystems. *Trends in Ecology & Evolution*, 20, 387–94.

Bond, W. J., Lee, W. G., & Craine, J. M. (2004). Plant structural defences against browsing birds: a legacy of New Zealand's extinct moas. *Oikos*, 104(3), 500–8.

Bond, W. J., & Loffell, D. (2001). Introduction of giraffe changes acacia distribution in a South African savanna. *African Journal of Ecology*, 39(3), 286–94.

Bond, W. J., & Midgley, J. J. (1995). Kill thy neighbour: an individualistic argument for the evolution of flammability. *Oikos*, 73(1), 79–85.

Bond, W. J., & Midgley, J. J. (2001). Ecology of sprouting in woody plants: the persistence niche. *Trends in Ecology & Evolution*, 16(1), 45–51.

Bond, W. J., & Midgley, J. J. (2012). Fire and the angiosperm revolutions. *International Journal of Plant Sciences*, 173(6), 569–83.

Bond, W. J., & Midgley, G. F. (2012). Carbon dioxide and the uneasy interactions of trees and savannah grasses. *Philosophical Transactions of the Royal Society of London B: Biological Sciences*, 367(1588), 601–12.

Bond, W. J., Midgley, G. F., and Woodward, F. I. (2003a). What controls South African vegetation—climate or fire? *South African Journal of Botany*, 69(1), pp. 79–91.

Bond, W.J., Midgley, G.F., Woodward, F.I. (2003b) The importance of low atmospheric CO_2 and fire in promoting the spread of grasslands and savannas. *Global Change Biology*, 9, 973–82.

Bond, W. J., & Parr, C. L. (2010). Beyond the forest edge: ecology, diversity and conservation of the grassy biomes. *Biological Conservation*, 143(10), 2395–404.

Bond, W. J., & Scott, A. C. (2010). Fire and the spread of flowering plants in the Cretaceous. *New Phytologist*, 188, 1137–50.

Bond, W. J., & Silander, J. A. (2007). Springs and wire plants: anachronistic defences against Madagascar's extinct elephant birds. *Proceedings of the Royal Society of London B: Biological Sciences*, 274(1621), 1985–92.

Bond, W. J., Silander Jr, J. A., Ranaivonasy, J., & Ratsirarson, J. (2008). The antiquity of Madagascar's grasslands and the rise of C4 grassy biomes. *Journal of Biogeography*, 35(10), 1743–58.

Bond, W. J., Smythe, K. A., & Balfour, D. A. (2001). Acacia species turnover in space and time in an African savanna. *Journal of Biogeography*, 28(1), 117–28.

Bond, W. J., & van Wilgen, B.W. (1996) *Fire and plants*. Chapman and Hall, London, UK.

Bond, W. J., Woodward, F. I., and Midgley, G. F. (2005). The global distribution of ecosystems in a world without fire. *New Phytologist*, 165, 525–38.

Bormann, B. T., Wang, D., Snyder, M. C., Bormann, F. H., Benoit, G., & April, R. (1998). Rapid, plant-induced weathering in an aggrading experimental ecosystem. *Biogeochemistry*, 43(2), 129–55.

Bosch, J. M., & Hewlett, J. D. (1982). A review of catchment experiments to determine the effect of vegetation changes on water yield and evapotranspiration. *Journal of Hydrology*, 55(1–4), 3–23.

Bouchenak-Khelladi, Y., Slingsby, J. A., Verboom, G. A., & Bond, W. J. (2014). Diversification of C4 grasses (Poaceae) does not coincide with their ecological dominance. *American Journal of Botany*, 101(2), 300–7.

Boucher, G. (1975). The *Orothamnus* saga. *Veld & Flora*, 61(2), 2–5.

Boundja, R. P., & Midgley, J. J. (2010). Patterns of elephant impact on woody plants in the Hluhluwe-Imfolozi park, Kwazulu-Natal, South Africa. *African Journal of Ecology*, 48(1), 206–14.

Bowman, D. M. (2000). *Australian rainforests: islands of green in a land of fire*. Cambridge University Press, Cambridge, UK.

Bowman, D. M., French, B. J., & Prior, L. D. (2014). Have plants evolved to self-immolate? *Frontiers in Plant Science*, 5, 590.

Bowman, D. M., Murphy, B. P., & Banfai, D. S. (2010). Has global environmental change caused monsoon rainforests to expand in the Australian monsoon tropics? *Landscape Ecology*, 25(8), 1247–60.

Bowman, D. M., Perry, G. L., Higgins, S. I., Johnson, C. N., Fuhlendorf, S. D., & Murphy, B. P. (2016). Pyrodiversity is the coupling of biodiversity and fire regimes in food webs. *Philosophical Transactions of the Royal Society of London B: Biological Sciences*, 371(1696), 20150169.

Bowman, D. M., & Wood, S. W. (2009). Fire-driven land cover change in Australia and WD Jackson's theory of the fire ecology of southwest Tasmania. In *Tropical Fire Ecology* (ed. M. Cochrane), pp. 87–111. Springer, Berlin, Heidelberg.

Box, E. O. (1981). *Macroclimate and Plant Forms: An Introduction to Predictive Modeling in Phytogeography*. Tasks for Vegetation Science, Vol. 1. Dr W. Junk BV. Publishers, The Hague 258 pp.

Box, E. O. (1996). Plant functional types and climate at the global scale. *Journal of Vegetation Science*, 7(3), 309–20.

Boy, J., & Wilcke, W. (2008). Tropical Andean forest derives calcium and magnesium from Saharan dust. *Global Biogeochemical Cycles*, 22(1), GB1027. doi:10.1029/2007GB002960

Boyce, C. K., Brodribb, T. J., Feild, T. S., & Zwieniecki, M. A. (2009). Angiosperm leaf vein evolution was physiologically and environmentally transformative. *Proceedings of the Royal Society of London B: Biological Sciences*, 276(1663), 1771–6.

Bradshaw, L., & Waller, D. M. (2016). Impacts of white-tailed deer on regional patterns of forest tree recruitment. *Forest Ecology and Management*, 375, 1–11.

Bradstock, R. A., Williams, R. J., & Gill, A. M. (eds) (2012). *Flammable Australia: fire regimes, biodiversity and ecosystems in a changing world*. CSIRO Publishing.

Bright, R. M., Zhao, K., Jackson, R. B., & Cherubini, F. (2015). Quantifying surface albedo and other direct biogeophysical climate forcings of forestry activities. *Global Change Biology*, 21(9), 3246–66.

Brodribb, T. J., & Feild, T. S. (2010). Leaf hydraulic evolution led a surge in leaf photosynthetic capacity during early angiosperm diversification. *Ecology Letters*, 13, 175–83.

Brooks, M. L., D'Antonio, C. M., Richardson, D. M., Grace, J. B., Keeley, J. E., DiTomaso, J. M., Hobbs, R. J., Pellant, M., & Pyke, D. (2004). Effects of invasive alien plants on fire regimes. *Bioscience*, 54(7), 677–88.

Brown, S. A., Scott, A. C., Glasspool, I. J., & Collinson, M. E. (2012). Cretaceous wildfires and their impact on the Earth system. *Cretaceous Research*, 36, 162–90.

Bucini, G., & Hanan, N. P. (2007). A continental-scale analysis of tree cover in African savannas. *Global Ecology and Biogeography*, 16(5), 593–605.

Buisson, E., Le Stradic, S., Silveira, F.A., Durigan, G., Overbeck, G.E., Fidelis, A., Fernandes, G.W., Bond, W.J., Hermann, J.M., Mahy, G., & Alvarado, S.T. (2018). Resilience and restoration of tropical and subtropical grasslands, savannas, and grassy woodlands. *Biological Reviews*. https://doi.org/10.1111/brv.12470

Buitenwerf, R., Bond, W. J., Stevens, N., & Trollope, W. S. W. (2012). Increased tree densities in South African

savannas: >50 years of data suggests CO_2 as a driver. *Global Change Biology*, 18(2), 675–84.

Burger, N., & Bond, W. J. (2015). Flammability traits of Cape shrubland species with different post-fire recruitment strategies. *South African Journal of Botany*, 101, 40–8.

Burney, D. A. (1987a). Late Holocene vegetational change in central Madagascar. *Quaternary Research*, 28, 130–43.

Burney, D. A. (1987b). Late Quaternary stratigraphic charcoal records from Madagascar. *Quaternary Research*, 28(2), 274–80.

Burney, D. A., Robinson, G. S., & Burney, L. P. (2003). *Sporormiella* and the late Holocene extinctions in Madagascar. *Proceedings of the National Academy of Sciences of the United States of America*, 100(19), 10800–5.

Burrows, G. E. (2002). Epicormic strand structure in *Angophora*, *Eucalyptus* and *Lophostemon* (Myrtaceae)—implications for fire resistance and recovery. *New Phytologist*, 153(1), 111–31.

Burrows, G. E., Hornby, S. K., Waters, D. A., Bellairs, S. M., Prior, L. D., & Bowman, D. M. J. S. (2010). A wide diversity of epicormic structures is present in Myrtaceae species in the northern Australian savanna biome–implications for adaptation to fire. *Australian Journal of Botany*, 58(6), 493–507.

Canadell, J., Jackson, R. B., Ehleringer, J. B., Mooney, H. A., Sala, O. E., & Schulze, E. D. (1996). Maximum rooting depth of vegetation types at the global scale. *Oecologia*, 108(4), 583–95.

Carpenter, R.J., Holman, A.I., Abell, A.D., and Grice, K. (2017). Cretaceous fire in Australia: a review with new geochemical evidence, and relevance to the rise of the angiosperms. *Australian Journal of Botany*, 64(8), 564–78.

Carpenter, R. J., Macphail, M. K., Jordan, G. J., & Hill, R. S. (2015). Fossil evidence for open, Proteaceae-dominated heathlands and fire in the Late Cretaceous of Australia. *American Journal of Botany*, 102(12), 2092–107.

Carpenter, S. R., & Kitchell, J. F. (1988). Consumer control of lake productivity. *BioScience*, 38(11), 764–9.

Carpenter, S. R., Kitchell, J. F., & Hodgson, J. R. (1985). Cascading trophic interactions and lake productivity. *BioScience*, 35(10), 634–9.

Carroll, S. B. (2016). *The Serengeti Rules: The quest to discover how life works and why it matters*. Princeton University Press, Princeton, NJ.

Cerling, T. E., Harris, J. M., MacFadden, B. J., Leakey, M. G., Quade, J., Eisenmann, V., & Ehleringer, J. R. (1997). Global vegetation change through the Miocene/Pliocene boundary. *Nature*, 389, 153–8.

Cerling, T. E., Harris, J. M., & Passey, B. H. (2003). Diets of East African Bovidae based on stable isotope analysis. *Journal of Mammalogy*, 84(2), 456–70.

Cerling, T. E., Wynn, J. G., Andanje, S. A., Bird, M. I., Korir, D. K., Levin, N. E., . . . & Remien, C. H. (2011). Woody cover and hominin environments in the past 6 million years. *Nature*, 476(7358), 51.

Chapin, F. S., Walker, L. R., Fastie, C. L., & Sharman, L. C. (1994). Mechanisms of primary succession following deglaciation at Glacier Bay, Alaska. *Ecological Monographs*, 64(2), 149–75.

Charles-Dominique, T., Barczi, J. F., Le Roux, E., & Chamaillé-Jammes, S. (2017b). The architectural design of trees protects them against large herbivores. *Functional Ecology*, 31(9), 1710–17.

Charles-Dominique, T., Beckett, H., Midgley, G. F., & Bond, W. J. (2015a). Bud protection: a key trait for species sorting in a forest–savanna mosaic. *New Phytologist*, 207(4), 1052–60.

Charles-Dominique, T., Davies, T. J., Hempson, G. P., Bezeng, B. S., Daru, B. H., Kabongo, R. M.,...& Bond, W. J. (2016). Spiny plants, mammal browsers, and the origin of African savannas. *Proceedings of the National Academy of Sciences of the United States of America*, 113(38), E5572–E5579.

Charles-Dominique, T., Midgley, G. F., & Bond, W. J. (2017a). Fire frequency filters species by bark traits in a savanna–forest mosaic. *Journal of Vegetation Science*, 28(4), 728–35.

Charles-Dominique, T., Midgley, G. F., Tomlinson, K. W., & Bond, W. J. (2018). Steal the light: shade vs fire adapted vegetation in forest–savanna mosaics. *New Phytologist*, 218(4), 1419–29.

Charles-Dominique, T., Staver, A. C., Midgley, G. F., & Bond, W. J. (2015b). Functional differentiation of biomes in an African savanna/forest mosaic. *South African Journal of Botany*, 101, 82–90.

Chazdon, R. L. (2017). Landscape restoration, natural regeneration, and the forests of the future. *Annals of the Missouri Botanical Garden*, 102(2), 251–7.

Chergui, B., Fahd, S., Santos, X., & Pausas, J. G. (2018). Socioeconomic factors drive fire-regime variability in the Mediterranean Basin. *Ecosystems*, 21(4), 619–28.

Clarke, P. J., Lawes, M. J., Midgley, J. J., Lamont, B. B., Ojeda, F., Burrows, G. E.,...& Knox, K. J. E. (2013). Resprouting as a key functional trait: how buds, protection and resources drive persistence after fire. *New Phytologist*, 197(1), 19–35.

Clements, F. E. (1936). Nature and structure of the climax. *Journal of Ecology*, 24(1), 252–84.

Cochrane, M. (ed.) (2009). *Tropical fire ecology: climate change, land use and ecosystem dynamics*. Springer Science & Business Media.

Cochrane, M. A. (2003). Fire science for rainforests. *Nature*, 421(6926), 913.

Cody, M. L. (1981). Habitat selection in birds: the roles of vegetation structure, competitors, and productivity. *BioScience*, 31(2), 107–13.

Cody, M. L., & Mooney, H. A. (1978). Convergence versus nonconvergence in Mediterranean-climate ecosystems. *Annual Review of Ecology and Systematics*, 9(1), 265–321.

Coetsee, C., Bond, W. J., & February, E. C. (2010). Frequent fire affects soil nitrogen and carbon in an African savanna by changing woody cover. *Oecologia*, 162(4), 1027–34.

Coetsee, C., Bond, W. J., & Wigley, B. J. (2015). Forest and fynbos are alternative states on the same nutrient poor geological substrate. *South African Journal of Botany*, 101, 57–65.

Cole, M. M. (1986). *The savannas: biogeography and geobotany*. Academic Press, London.

Collatz, G. J., Berry, J. A., & Clark, J. S. (1998). Effects of climate and atmospheric CO_2 partial pressure on the global distribution of C_4 grasses: present, past, and future. *Oecologia*, 114(4), 441–54.

Collinson, M. E., Featherstone, C., Cripps, J. A., Nichols, G. J., & Scott, A. C. (2000). Charcoal-rich plant debris accumulations in the lower Cretaceous of the Isle of Wight, England. *Acta Palaeobotanica*, 2(suppl), 93–105.

Compton, R. H. (1926). Veld burning and veld deterioration. *S. Afr. J. Nat. Hist.*, 6 (1), 5–19.

Connell, J. H., & Slatyer, R. O. (1977). Mechanisms of succession in natural communities and their role in community stability and organization. *The American Naturalist*, 111(982), 1119–44.

Cook, G. D., & Goyens, C. M. (2008). The impact of wind on trees in Australian tropical savannas: lessons from Cyclone Monica. *Austral Ecology*, 33(4), 462–70.

Cooper, S. M., & Owen-Smith, N. (1986). Effects of plant spinescence on large mammalian herbivores. *Oecologia*, 68(3), 446–55.

Coughenour, M. B. (1985). Graminoid responses to grazing by large herbivores: adaptations, exaptations, and interacting processes. *Annals of the Missouri Botanical Garden*, 72(4), 852–63.

Côté, S. D., Rooney, T. P., Tremblay, J. P., Dussault, C., & Waller, D. M. (2004). Ecological impacts of deer overabundance. *Annu. Rev. Ecol. Evol. Syst.*, 35, 113–47.

Cowling, R. M. (1992). *The ecology of fynbos. Nutrients, fire and diversity*. Cape Town.: Oxford Univ. Press. Contents include: Flora and vegetation, by RM Cowling & PM Holmes, 23–61.

Cowling, R. M., & Campbell, B. M. (1980). Convergence in vegetation structure in the Mediterranean communities of California, Chile, and South Africa. *Vegetatio*, 43(3), 191–7.

Cowling, R. M., Rundel, P. W., Lamont, B. B., Kalin Arroyo, M., & Arianoutsou, M. (1996). Plant diversity in Mediterranean-climate regions. *Trends in Ecology & Evolution*, 11(9), 362–66.

Craine, J. M. (2009). *Resource strategies of wild plants*. Princeton University Press, Princeton, NJ.

Cramer, M.D. (2010). Phosphate as a limiting resource: introduction. *Plant and Soil*, 334 (1), 1–10.

Cramer, M. D., Hawkins, H. J., & Verboom, G. A. (2009). The importance of nutritional regulation of plant water flux. *Oecologia*, 161(1), 15–24.

Cramer, M. D., Power, S. C., Belev, A., Gillson, L., Bond, W. J., Hoffman, M. T., & Hedin, L. O. (2019). Are forest–shrubland mosaics of the Cape Floristic Region an example of alternate stable states? *Ecography*, 42 (4), 717–29. https://doi.org/10.1111/ecog.03860

Cramer, M., West, A., Power, S., Skelton, R., & Stock, W. D. (2014). Plant ecophysiological diversity. In: N. Allsopp, J.F. Colville, G.A. Verboom (eds), *Fynbos: Ecology, evolution, and conservation of a megadiverse region*, Oxford University Press. Pp. 248–72.

Cramer, W., Bondeau, A., Woodward, F. I., Prentice, I. C., Betts, R. A., Brovkin, V., Cox, P. M., Fisher, V., Foley, J. A., Friend, A. D., & Kucharik, C. (2001). Global response of terrestrial ecosystem structure and function to CO_2 and climate change: results from six dynamic global vegetation models. *Global Change Biology*, 7(4), 357–73.

Cramer, W. P., & Leemans, R. (1993). Assessing impacts of climate change on vegetation using climate classification systems. In Vegetation dynamics & global change (pp. 190–217). Springer US.

Crane, P. R., & Lidgard, S. (1989). Angiosperm diversification and paleolatitudinal gradients in Cretaceous floristic diversity. *Science*, 246, 675–8.

Crisp, M. D., Arroyo, M. T., Cook, L. G., Gandolfo, M. A., Jordan, G. J., McGlone, M. S.,…& Linder, H. P. (2009). Phylogenetic biome conservatism on a global scale. *Nature*, 458(7239), 754.

Cromsigt, J. P., & te Beest, M. (2014). Restoration of a megaherbivore: landscape-level impacts of white rhinoceros in Kruger National Park, South Africa. *Journal of Ecology*, 102(3), 566–75.

Cromsigt, J. P., Michiel, P., Veldhuis, W. D., Le Roux, E., Gosling, C. M., & Archibald, S. (2017). The functional ecology of grazing lawns: how grazers, termites, people, and fire shape HiP's savanna grassland mosaic. *Conserving Africa's Mega-Diversity in the Anthropocene: The Hluhluwe-iMfolozi Park Story*, 135. by Joris P. G. M. Cromsigt (Editor), Sally Archibald (Editor), Norman Owen-Smith (Editor). Cambridge University Press.

Dantas, V. D. L., Hirota, M., Oliveira, R. S., & Pausas, J. G. (2016). Disturbance maintains alternative biome states. *Ecology Letters*, 19(1), 12–19.

D'Antonio, C. M., Hughes, R. F., & Tunison, J. T. (2011). Long-term impacts of invasive grasses and subsequent fire in seasonally dry Hawaiian woodlands. *Ecological Applications*, 21(5), 1617–28.

D'Antonio, C. M., & Vitousek, P. M. (1992). Biological invasions by exotic grasses, the grass/fire cycle, and global change. *Annual Review of Ecology and Systematics*, 23(1), 63–87.

Darwin, C. (1840). *Journal of Researches Into the Geology and Natural History of the Various Countries Visited by HMS Beagle, Under the Command of Captain Fitzroy from 1832 to 1836*. Colburn, London.

Daskin, J. H., Stalmans, M., & Pringle, R. M. (2016). Ecological legacies of civil war: 35-year increase in savanna tree cover following wholesale large-mammal declines. *Journal of Ecology*, 104(1), 79–89.

Dass, P., Houlton, B. Z., Wang, Y., & Warlind, D. (2018). Grasslands may be more reliable carbon sinks than forests in California. *Environmental Research Letters*, 13(7), 074027.

Davis, C. C., Webb, C. O., Wurdack, K. J., Jaramillo, C. A., & Donoghue, M. J. (2005). Explosive radiation of Malpighiales supports a MidCretaceous origin of modern tropical rain forests. *American Naturalist*, 165, E36–E65.

Devine, A. P., McDonald, R. A., Quaife, T., & Maclean, I. M. (2017). Determinants of woody encroachment and cover in African savannas. *Oecologia*, 183(4), 939–951.

Diamond, J. M. (1975). The island dilemma: lessons of modern biogeographic studies for the design of natural reserves. *Biological Conservation*, 7(2), 129–46.

Díaz, S., Hodgson, J. G., Thompson, K., Cabido, M., Cornelissen, J. H. C., Jalili, A.,…& Band, S. R. (2004). The plant traits that drive ecosystems: evidence from three continents. *Journal of Vegetation Science*, 15(3), 295–304.

Díaz, S., Kattge, J., Cornelissen, J. H., Wright, I. J., Lavorel, S., Dray, S.,…& Garnier, E. (2016). The global spectrum of plant form and function. *Nature*, 529(7585), 167.

Díaz, S., Lavorel, S., McIntyre, S., Falczuk, V., Casanoves, F., Milchunas, D. G.,…& Landsberg, J. (2007). Plant trait responses to grazing–a global synthesis. *Global Change Biology*, 13(2), 313–341.

Dinerstein, E., Olson, D., Joshi, A., Vynne, C., Burgess, N. D., Wikramanayake, E., Hahn, N., Palminteri, S., Hedao, P., Noss, R. & Hansen, M. (2017). An ecoregion-based approach to protecting half the terrestrial realm. *BioScience*, 67(6), 534–45.

Donaldson, J. E., Archibald, S., Govender, N., Pollard, D., Luhdo, Z., & Parr, C. L. (2018). Ecological engineering through fire–herbivory feedbacks drives the formation of savanna grazing lawns. *Journal of Applied Ecology*, 55(1), 225–35.

Doughty, C. E., Faurby, S., & Svenning, J. C. (2016a). The impact of the megafauna extinctions on savanna woody cover in South America. *Ecography*, 39(2), 213–222.

Doughty, C. E., Roman, J., Faurby, S., Wolf, A., Haque, A., Bakker, E. S.,…& Svenning, J. C. (2016b). Global nutrient transport in a world of giants. *Proceedings of the National Academy of Sciences of the United States of America*, 113(4), 868–873.

Dubinin, M., Luschekina, A., & Radeloff, V. C. (2011). Climate, livestock, and vegetation: what drives fire increase in the arid ecosystems of southern Russia? *Ecosystems*, 14(4), 547–562.

Dunn, R. E., Strömberg, C. A., Madden, R. H., Kohn, M. J., & Carlini, A. A. (2015). Linked canopy, climate, and faunal change in the Cenozoic of Patagonia. *Science*, 347(6219), 258–61.

Dupont, L. M., Jahns, S., Marret, F., & Ning, S. (2000). Vegetation change in equatorial West Africa: time-slices for the last 150 ka. *Palaeogeography, Palaeoclimatology, Palaeoecology*, 155(1), 95–122.

Durigan, G., & Ratter, J. A. (2006). Successional changes in cerrado and cerrado/forest ecotonal vegetation in western Sao Paulo State, Brazil, 1962–2000. *Edinburgh Journal of Botany*, 63(1), 119–130.

Durigan, G., & Ratter, J. A. (2016). The need for a consistent fire policy for Cerrado conservation. *Journal of Applied Ecology*, 53(1), 11–15.

du Toit, J. T. (1990). Feeding-height stratification among African browsing ruminants. *African Journal of Ecology*, 28(1), 55–61.

Eamus, D., Myers, B., Duff, G., & Williams, R.J. (1999). Seasonal changes in photosynthesis of eight savanna tree species. *Tree Physiol.*, 19, 665–71.

Edwards, E. J., Osborne, C. P., Strömberg, C. A. E., Smith, S. A., & the C4 Grasses Consortium (2010). The origins of C4 grasslands: integrating evolutionary and ecosystem science. *Science*, 328, 587–91.

Ehleringer, J. R., Cerling, T. E., Helliker, B. R. (1997). C4 photosynthesis, atmospheric CO_2, and climate. *Oecologia*, 112, 285–99.

Ehrenfeld, J. G., Ravit, B., & Elgersma, K. (2005). Feedback in the plant-soil system. *Annual Review of Environment and Resources*, 30, 75–115.

Elith, J. and Leathwick, J.R. (2009). Species distribution models: ecological explanation and prediction across space and time. *Annual review of ecology, evolution, and systematics*, 40, pp. 677–97.

Erdős, L., Ambarlı, D., Anenkhonov, O. A., Bátori, Z., Cserhalmi, D., Kiss, M., ... & Török, P. (2018). The edge of two worlds: A new review and synthesis on Eurasian forest-steppes. *Applied Vegetation Science*, 21(3), 345–62.

Erwin, D. H. (2008). Macroevolution of ecosystem engineering, niche construction and diversity. *Trends in Ecology & Evolution*, 23(6), 304–10.

Everson, C. S., Everson, T. M., & Tainton, N. M. (1988). Effects of intensity and height of shading on the tiller initiation of six grass species from the Highland sourveld of Natal. *South African Journal of Botany*, 54(4), 315–18.

Fairhead, J., & Leach, M. (1995). False forest history, complicit social analysis: rethinking some West African environmental narratives. *World Development*, 23(6), 1023–35.

Falcon-Lang, H. J. (2000). Fire ecology of the Carboniferous tropical zone. *Palaeogeography, Palaeoclimatology, Palaeoecology*, 164, 339–55.

Fan, Y., Miguez-Macho, G., Jobbágy, E. G., Jackson, R. B., & Otero-Casal, C. (2017). Hydrologic regulation of plant rooting depth. *Proceedings of the National Academy of Sciences of the United States of America*, 114 (40), 10572–7.

Farley, K. A., Jobbágy, E. G., & Jackson, R. B. (2005). Effects of afforestation on water yield: a global synthesis with implications for policy. *Global Change Biology*, 11(10), 1565–76.

February, E. C., Higgins, S. I., Bond, W. J., & Swemmer, L. (2013). Influence of competition and rainfall manipulation on the growth responses of savanna trees and grasses. *Ecology*, 94(5), 1155–64.

Feild, T. S., & Arens, N. C. (2005). Form, function and environments of the early angiosperms: merging extant phylogeny and ecophysiology with fossils. *New Phytologist*, 166(2), 383–408.

Feild, T. S., Brodribb, T. J., Iglesias, A., Chatelet, D. S., Baresch, A., Upchurch, Jr, G. R., Gomez, B., et al. (2011). Fossil evidence for Cretaceous escalation in angiosperm leaf vein evolution. *Proceedings of the National Academy of Sciences of the United States of America*, 108, 8363–6.

Fensham, R. J., & Bowman, D. M. J. S. (1992). Stand structure and the influence of overwood on regeneration in tropical eucalypt forest on Melville Island. *Aust. J. Bot.*, 40, 335–52.

Fensham, R. J., Fairfax, R. J., & Ward, D. P. (2009). Drought-induced tree death in savanna. *Global Change Biology*, 15(2), 380–7.

Fensham, R. J., & Holman, J. E. (1999). Temporal and spatial patterns in drought-related tree dieback in Australian savanna. *Journal of Applied Ecology*, 36(6), 1035–50.

Fenton, J. H. (2008). A postulated natural origin for the open landscape of upland Scotland. *Plant Ecology & Diversity*, 1(1), 115–27.

Fernandes, G. W. (ed.) (2016). *Ecology and conservation of mountaintop grasslands in Brazil*. Springer, Cham, Switzerland.

Fidelis, A., & Blanco, C. (2014). Does fire induce flowering in Brazilian subtropical grasslands? *Applied Vegetation Science*, 17(4), 690–9.

Fisher, B.L., & Robertson, H.G. (2002) Comparison and origin of forest and grassland ant assemblages in the high plateau of Madagascar (Hymenoptera: Formicidae). *Biotropica*, 34, 155–67.

Fisher, J. L., Loneragan, W. A., Dixon, K., Delaney, J., & Veneklaas, E. J. (2009). Altered vegetation structure and composition linked to fire frequency and plant invasion in a biodiverse woodland. *Biological Conservation*, 142(10), 2270–81.

Flannery, T. (1994). *The future eaters*. Chatswood. New South Wales: Reed Books.

Flannigan, M., Cantin, A. S., De Groot, W. J., Wotton, M., Newbery, A., & Gowman, L. M. (2013). Global wildland fire season severity in the 21st century. *Forest Ecology and Management*, 294, 54–61.

Fölster, H., Dezzeo, N., & Priess, J. A. (2001). Soil–vegetation relationship in base-deficient premontane moist forest–savanna mosaics of the Venezuelan Guayana. *Geoderma*, 104(1–2), 95–113.

Franczak, D. D., Marimon, B. S., Hur Marimon-Junior, B., Mews, H. A., Maracahipes, L., & Oliveira, E. A. D. (2011). Changes in the structure of a savanna forest over a six-year period in the Amazon-Cerrado transition, Mato Grosso state, Brazil. *Rodriguésia*, 62(2), 425–36.

Frank, D. A., McNaughton, S. J., & Tracy, B. F. (1998). The ecology of the earth's grazing ecosystems. *BioScience*, 48(7), 513–21.

Friis, E. M., Crane, P. R., Pedersen, K. R. (2011). *Early flowers and angiosperm evolution*. Cambridge University Press, Cambridge, UK.

Fuhlendorf, S. D., Engle, D. M., Kerby, J. A. Y., & Hamilton, R. (2009). Pyric herbivory: rewilding landscapes through the recoupling of fire and grazing. *Conservation Biology*, 23(3), 588–98.

Fuhlendorf, S. D., Harrell, W. C., Engle, D. M., Hamilton, R. G., Davis, C. A., & Leslie, D. M. (2006). Should heterogeneity be the basis for conservation? Grassland bird response to fire and grazing. *Ecological Applications*, 16(5), 1706–16.

Fyfe, R. M., Twiddle, C., Sugita, S., Gaillard, M. J., Barratt, P., Caseldine, C. J., . . . & Grant, M. J. (2013). The Holocene vegetation cover of Britain and Ireland: overcoming problems of scale and discerning patterns of openness. *Quaternary Science Reviews*, 73, 132–48.

Gibbs Russell, G. E. (1987). Preliminary floristic analysis of the major biomes in southern Africa. *Bothalia*, 17(2), 213–27.

Giglio, L., Randerson, J. T., & van der Werf, G. R. (2013). Analysis of daily, monthly, and annual burned area using the fourth-generation global fire emissions database (GFED4). *Journal of Geophysical Research: Biogeosciences*, 118(1), 317–28.

Gill, A. M. (1975). Fire and the Australian flora: a review. *Australian Forestry*, 38(1), 4–25.

Gill, J. L. (2014). Ecological impacts of the late Quaternary megaherbivore extinctions. *New Phytologist*, 201(4), 1163–9.

Gill, J. L., Williams, J. W., Jackson, S. T., Lininger, K. B., & Robinson, G. S. (2009). Pleistocene megafaunal collapse, novel plant communities, and enhanced fire regimes in North America. *Science*, 326(5956), 1100–3.

Gilliam, F. S., & Platt, W. J. (1999). Effects of long-term fire exclusion on tree species composition and stand structure in an old-growth *Pinus palustris* (longleaf pine) forest. *Plant Ecology*, 140(1), 15–26.

Gillson, L. (2015). Evidence of a tipping point in a southern African savanna?. *Ecological Complexity*, 21, 78–86.

Glasspool, I. J., & Scott, A. C. (2010). Phanerozoic concentrations of atmospheric oxygen reconstructed from sedimentary charcoal. *Nature Geoscience*, 3(9), 627.

Godfrey, L. R., & Crowley, B. E. (2016). Madagascar's ephemeral palaeo-grazer guild: who ate the ancient C4 grasses?. *Proceedings of the Royal Society of London B: Biological Sciences*, 283(1834), 20160360.

Goldblatt, P., & Manning, J. C. (2002). Plant diversity of the Cape region of southern Africa. *Annals of the Missouri Botanical Garden*, 89(2), 281–302.

Goodland, R., & Pollard, R. (1973). The Brazilian cerrado vegetation: a fertility gradient. *Journal of Ecology*, 61, 219–24.

Gowda, J. H. (1996). Spines of *Acacia tortilis*: what do they defend and how? *Oikos*, 77(2), 279–84.

Grandidier, A. (1898). Le boisement de l'Imerina. *Bulletin du Comité de Madagascar*, 4, 83–7.

Gray, E. F., & Bond, W. J. (2015). Soil nutrients in an African forest/savanna mosaic: drivers or driven? *South African Journal of Botany*, 101, 66–72.

Green, G.M. & Sussman, R.W. (1990). Deforestation history of the eastern rain forests of Madagascar from satellite images. *Science*, 248, 212–15.

Greenwood, R. M., & Atkinson, I. A. E. (1977). Evolution of divaricating plants in New Zealand in relation to moa browsing. In *Proceedings (New Zealand Ecological Society)* (pp. 21–33). New Zealand Ecological Society (Inc.).

Greve, M., Reyers, B., Lykke, A. M., & Svenning, J. C. (2013). Spatial optimization of carbon-stocking projects across Africa integrating stocking potential with co-benefits and feasibility. *Nature Communications*, 4, 2975.

Grime, J. P. (1979). *Plant strategies and vegetation processes*. John Wiley & Sons, Chichester, UK.

Grime, J. P. (2006). *Plant strategies, vegetation processes, and ecosystem properties*. 2nd edition. John Wiley & Sons, Chichester, UK.

Grisebach, A. (1872). Die Vegetation der Erdenach ihrer klimatischen. *Anordnung Bd. I und II. Leipzig*.

Hackel, J., Vorontsova, M. S., Nanjarisoa, O. P., Hall, R. C., Razanatsoa, J., Malakasi, P., & Besnard, G. (2018). Grass diversification in Madagascar: *in situ* radiation of two large C3 shade clades and support for a Miocene to Pliocene origin of C4 grassy biomes. *Journal of Biogeography*, 45(4), 750–61.

Hagen-Thorn, A., Callesen, I., Armolaitis, K., & Nihlgård, B. (2004). The impact of six European tree species on the chemistry of mineral topsoil in forest plantations on former agricultural land. *Forest Ecology and Management*, 195(3), 373–84.

Hairston, N. G., Smith, F. E., & Slobodkin, L. B. (1960). Community structure, population control, and competition. *American Naturalist*, 94(879), 421–5.

Hanan, N. P., Tredennick, A. T., Prihodko, L., Bucini, G., & Dohn, J. (2014). Analysis of stable states in global savannas: is the CART pulling the horse? *Global Ecology and Biogeography*, 23(3), 259–63.

Hansen, M. C., Potapov, P. V., Moore, R., Hancher, M., Turubanova, S. A. A., Tyukavina, A.,…& Kommareddy, A. (2013). High-resolution global maps of 21st-century forest cover change. *Science*, 342(6160), 850–3.

Hantson, S., Arneth, A., Harrison, S. P., Kelley, D. I., Prentice, I. C., Rabin, S. S.,…& Bachelet, D. (2016). The status and challenge of global fire modelling. *Biogeosciences*, 13(11), 3359–75.

Haridasan, M. (1992). Observations on soils, foliar nutrient concentrations and floristic composition of cerrado sensu stricto and cerradão communities in central Brazil. In: Furley, P., Ratter, J., Proctor, J. (Eds.) *Nature and dynamics of forest–savanna boundaries*. Chapman & Hall, London, UK.

Haxeltine, A., & Prentice, I. C. (1996). BIOME3: An equilibrium terrestrial biosphere model based on ecophysiological constraints, resource availability and competition among plant functional types. *Global Biogeochemical Cycles*, 10(4), 693–709.

Hazeldine, A., & Kirkpatrick, J. B. (2015). Practical and theoretical implications of a browsing cascade in Tasmanian forest and woodland. *Australian Journal of Botany*, 63(5), 435–43.

He, T., Lamont, B. B., & Downes, K. S. (2011). Banksia born to burn. *New Phytologist*, 191(1), 184–96.

He, T., Lamont, B. B., & Manning, J. (2016). A Cretaceous origin for fire adaptations in the Cape flora. *Scientific Reports*, 6, 34880.

He, T., Pausas, J. G., Belcher, C. M., Schwilk, D. W., & Lamont, B. B. (2012). Fire-adapted traits of *Pinus* arose in the fiery Cretaceous. *New Phytologist*, 194(3), 751–9.

Heikkinen, R. K., Luoto, M., Araújo, M. B., Virkkala, R., Thuiller, W., & Sykes, M. T. (2006). Methods and uncertainties in bioclimatic envelope modelling under climate change. Progress in Physical Geography, 30(6), 751–77.

Hempson, G. P., Archibald, S., & Bond, W. J. (2015a). A continent-wide assessment of the form and intensity of large mammal herbivory in Africa. *Science*, 350(6264), 1056–61.

Hempson, G. P., Archibald, S., & Bond, W. J. (2017). The consequences of replacing wildlife with livestock in Africa. *Scientific Reports*, 7(1), 17196.

Hempson, G. P., Archibald, S., Bond, W. J., Ellis, R. P., Grant, C. C., Kruger, F. J., Kruger, L. M., Moxley, C., Owen-Smith, N., Peel, M. J., and Smit, I. P. (2015b). Ecology of grazing lawns in Africa. *Biological Reviews*, 90(3), pp. 979–94.

Herring, J. R. (1985). Charcoal fluxes into sediments of the North Pacific Ocean: the Cenozoic record of burning. The carbon cycle and atmospheric CO2: natural variations Archean to present. *Chapman conference papers*, 32, 419–42. 10.1029/GM032p0419

Hidding, B., Tremblay, J. P., & Côté, S. D. (2013). A large herbivore triggers alternative successional trajectories in the boreal forest. *Ecology*, 94(12), 2852–60.

Higgins, S. I., Bond, W. J., & Trollope, W. S. (2000). Fire, resprouting and variability: a recipe for grass–tree coexistence in savanna. *Journal of Ecology*, 88(2), 213–29.

Higgins, S. I., Bond, W. J., February, E. C., Bronn, A., Euston-Brown, D. I., Enslin, B., Govender, N., Rademan, L., O'Regan, S., Potgieter, A. L., and Scheiter, S. (2007). Effects of four decades of fire manipulation on woody vegetation structure in savanna. *Ecology*, 88(5), pp. 1119–25.

Higgins, S. I., & Scheiter, S. (2012). Atmospheric CO_2 forces abrupt vegetation shifts locally, but not globally. *Nature*, 488(7410), 209.

Hirota, M., Holmgren, M., Van Nes, E. H., & Scheffer, M. (2011). Global resilience of tropical forest and savanna to critical transitions. *Science*, 334(6053), 232–5.

Hoetzel, S., Dupont, L., Schefuß, E., Rommerskirchen, F., & Wefer, G. (2013). The role of fire in Miocene to Pliocene C4 grassland and ecosystem evolution. *Nature Geoscience*, 6(12), 1027.

Hoffmann, W. A., Bazzaz, F. A., Chatterton, N. J., Harrison, P. A., & Jackson, R. B. (2000). Elevated CO_2 enhances resprouting of a tropical savanna tree. *Oecologia*, 123(3), 312–17.

Hoffmann, W. A., Geiger, E. L., Gotsch, S. G., Rossatto, D. R., Silva, L. C., Lau, O. L., Haridasan, M., & Franco, A. C. (2012a). Ecological thresholds at the savanna–forest boundary: how plant traits, resources and fire govern the distribution of tropical biomes. *Ecology Letters*, 15(7), 759–68.

Hoffmann, W. A., Jaconi, S. Y., Mckinley, K. L., Geiger, E. L., Gotsch, S. G., & Franco, A. C. (2012b). Fuels or microclimate? Understanding the drivers of fire feedbacks at savanna–forest boundaries. Austral Ecology, 37(6), 634–43.

Hoffmann, W. A., Orthen, B., & Do Nascimento, P. K. V. (2003). Comparative fire ecology of tropical savanna and forest trees. *Functional Ecology*, 17(6), 720–6.

Hoffmann, W. A., Orthen, B., & Franco, A. C. (2004). Constraints to seedling success of savanna and forest trees across the savanna–forest boundary. *Oecologia*, 140(2), 252–60.

Hoffmann, V., Verboom, G. A., & Cotterill, F. P. (2015). Dated plant phylogenies resolve Neogene climate and landscape evolution in the Cape Floristic Region. *PLoS One*, 10(9), e0137847.

Hofmeyr, S. D., Symes, C. T., & Underhill, L. G. (2014). Secretary bird *Sagittarius serpentarius* population trends and ecology: insights from South African citizen science data. *PloS One*, 9(5), e96772.

Holdo, R. M. (2007). Elephants, fire, and frost can determine community structure and composition in Kalahari woodlands. *Ecological Applications*, 17(2), 558–68.

Holdo, R. M., Mack, M. C., & Arnold, S. G. (2012). Tree canopies explain fire effects on soil nitrogen, phosphorus and carbon in a savanna ecosystem. *Journal of Vegetation Science*, 23(2), 352–60.

Holdo, R. M., Sinclair, A. R., Dobson, A. P., Metzger, K. L., Bolker, B. M., Ritchie, M. E., & Holt, R. D. (2009). A disease-mediated trophic cascade in the Serengeti and its implications for ecosystem C. *PLoS Biology*, 7(9), e1000210.

Holdridge, L. R. (1947). Determination of World Plant Formations from Simple Climatic Data, Science 105, 367–8.

Honda, E. A., & Durigan, G. (2016). Woody encroachment and its consequences on hydrological processes in the savannah. *Philosophical Transactions of the Royal Society of London B: Biological Sciences*, 371(1703), 20150313.

Hopper, S. D. (2009). OCBIL theory: towards an integrated understanding of the evolution, ecology and conservation of biodiversity on old, climatically buffered, infertile landscapes. *Plant and Soil*, 322(1–2), 49–86.

Hopper, S. D., & Gioia, P. (2004). The southwest Australian floristic region: evolution and conservation of a global hot spot of biodiversity. *Annu. Rev. Ecol. Evol. Syst.*, 35, 623–50.

Huber, O. (2006). Herbaceous ecosystems on the Guayana Shield, a regional overview. *Journal of Biogeography*, 33(3), 464–75.

Jackson, R. B., Canadell, J., Ehleringer, J. R., Mooney, H. A., Sala, O. E., & Schulze, E. D. (1996). A global analysis of root distributions for terrestrial biomes. *Oecologia*, 108(3), 389–411.

Jackson, R. B., Jobbágy, E. G., Avissar, R., Roy, S. B., Barrett, D. J., Cook, C. W., Farley, K. A., Le Maitre, D. C., McCarl, B. A., & Murray, B. C. (2005). Trading water for carbon with biological carbon sequestration. *Science*, 310(5756), 1944–7.

Jackson, W. D. (1968). Fire, air, water and earth–an elemental ecology of Tasmania. *Proceedings of the Ecological Society of Australia*, 3(9), 16).

Jacobs, B. F., Kingston, J. D., & Jacobs, L. L. (1999). The origin of grass-dominated ecosystems. *Annals of the Missouri Botanical Garden*, 86(2), 590–643.

Janzen, D. H., & Martin, P. S. (1982). Neotropical anachronisms: the fruits the gomphotheres ate. *Science*, 215(4528), 19–27.

Jeffery, K. J., Korte, L., Palla, F., Walters, G. M., White, L., & Abernethy, K. (2014). Fire management in a changing landscape: a case study from Lopé National Park, Gabon. *PARKS. The International Journal of Protected Areas and Conservation*, 20(1), 39–52.

Jobbágy, E. G., & Jackson, R. B. (2001). The distribution of soil nutrients with depth: global patterns and the imprint of plants. *Biogeochemistry*, 53(1), 51–77.

Jobbágy, E. G., & Jackson, R. B. (2004). The uplift of soil nutrients by plants: biogeochemical consequences across scales. *Ecology*, 85(9), 2380–9.

Johnson, C. (2006). *Australia's mammal extinctions: a 50,000-year history*. Cambridge University Press, Cambridge, UK.

Johnson, C. N., Prior, L. D., Archibald, S., Poulos, H. M., Barton, A. M., Williamson, G. J., & Bowman, D. M. (2018). Can trophic rewilding reduce the impact of fire in a more flammable world? *Philosophical Transactions of the Royal Society of London B: Biological Sciences*, 373(1761), 20170443.

Jones, C. G., Lawton, J. H., & Shachak, M. (1994). Organisms as ecosystem engineers. *Oikos*, 69(3), 373–86.

Jordan, G. J., Carpenter, R. J., & Brodribb, T. J. (2014). Using fossil leaves as evidence for open vegetation. *Palaeogeography, Palaeoclimatology, Palaeoecology*, 395, 168–75.

Jurena, P. N., & Archer, S. (2003). Woody plant establishment and spatial heterogeneity in grasslands. *Ecology*, 84(4), 907–19.

Keddy, P. (2007). *Plants and vegetation: origins, processes, consequences*. Cambridge University Press, Cambridge, UK.

Keeley, J. E., Bond, W. J., Bradstock, R. A., Pausas, J. G., & Rundel, P. W. (2012). *Fire in Mediterranean ecosystems: ecology, evolution and management*. Cambridge University Press, Cambridge, UK.

Keeley, J. E., & Rundel, P. W. (2005). Fire and the Miocene expansion of C4 grasslands. *Ecology Letters*, 8, 683–90.

Keeling, H. C., & Phillips, O. L. (2007). The global relationship between forest productivity and biomass. *Global Ecology and Biogeography*, 16(5), 618–31.

Kellman, M. (1984). Synergistic relationships between fire and low soil fertility in neotropical savannas: a hypothesis. *Biotropica*, 16(2), 158–60.

Kgope, B. S., Bond, W. J., & Midgley, G. F. (2010). Growth responses of African savanna trees implicate atmospheric $[CO_2]$ as a driver of past and current changes in savanna tree cover. *Austral Ecology*, 35(4), 451–63.

King, D. A. (1997). The functional significance of leaf angle in *Eucalyptus*. *Australian Journal of Botany*, 45(4), 619–39.

Kitayama, K., & Aiba, S. I. (2002). Ecosystem structure and productivity of tropical rain forests along altitudinal gradients with contrasting soil phosphorus pools on Mount Kinabalu, Borneo. *Journal of Ecology*, 90(1), 37–51.

Koechlin, J. (1993). Grasslands of Madagascar. Natural grasslands: eastern hemisphere and resume (ed. by R.T. Coupland), Ecosystems of the world 8B, pp. 291–301. Elsevier, Amsterdam.

Kooyman, R. M., Laffan, S. W., & Westoby, M. (2017). The incidence of low phosphorus soils in Australia. *Plant and Soil*, 412(1–2), 143–50.

Kreuter, U. P., Woodard, J. B., Taylor, C. A., & Teague, W. R. (2008). Perceptions of Texas landowners regarding fire and its use. *Rangeland Ecology & Management*, 61(4), 456–64.

Krook, K., Bond, W. J., & Hockey, P. A. (2007). The effect of grassland shifts on the avifauna of a South African savanna. *Ostrich: Journal of African Ornithology*, 78(2), 271–9.

Kruess, A., & Tscharntke, T. (2002). Grazing intensity and the diversity of grasshoppers, butterflies, and trap-nesting bees and wasps. *Conservation Biology*, 16(6), 1570–80.

Kruger, F. J., & Taylor, H. C. (1980). Plant species diversity in Cape fynbos: gamma and delta diversity. *Vegetatio*, 41(2), 85–93.

Kull, C.A. (2000). Deforestation, erosion and fire: degradation myths in the environmental history of Madagascar. *Environment and History*, 6, 423–50.

Kull, C.A. (2004). *Isle of fire: the political ecology of landscape burning in Madagascar*. The University of Chicago Press, Chicago.

Kulmatiski, A. (2018). Community-level plant–soil feed-backs explain landscape distribution of native and non-native plants. *Ecology and Evolution*, 8(4), 2041–9.

Lambers, H., Raven, J. A., Shaver, G. R., & Smith, S. E. (2008). Plant nutrient-acquisition strategies change with soil age. *Trends in Ecology & Evolution*, 23(2), 95–103.

Lamont, B. B., & He, T. (2012). Fire-adapted Gondwanan angiosperm floras evolved in the Cretaceous. *BMC Evolutionary Biology*, 12(1), 223.

Lapola, D. M., Martinelli, L. A., Peres, C. A., Ometto, J. P., Ferreira, M. E., Nobre, C. A., . . . & Joly, C. A. (2014). Pervasive transition of the Brazilian land-use system. *Nature Climate Change*, 4(1), 27.

Laris, P., & Wardell, D. A. (2006). Good, bad or 'necessary evil'? Reinterpreting the colonial burning experiments in the savanna landscapes of West Africa. *Geographical Journal*, 172(4), 271–90.

Laurance, W. F., Fearnside, P. M., Laurance, S. G., Delamonica, P., Lovejoy, T. E., Rankin-de Merona, J. M., . . . & Gascon, C. (1999). Relationship between soils and Amazon forest biomass: a landscape-scale study. *Forest Ecology and Management*, 118(1–3), 127–38.

Lawes, M. J., Midgley, J. J., & Clarke, P. J. (2013). Costs and benefits of relative bark thickness in relation to fire damage: a savanna/forest contrast. *Journal of Ecology*, 101(2), 517–24.

Le Roux, E., Kerley, G. I., & Cromsigt, J. P. (2018). Megaherbivores modify trophic cascades triggered by fear of predation in an African savanna ecosystem. *Current Biology*, 28(15), 2493–9.

Lehmann, C. E., Archibald, S. A., Hoffmann, W. A., & Bond, W. J. (2011). Deciphering the distribution of the savanna biome. *New Phytologist*, 191(1), 197–209.

Lehmann, C. E., Anderson, T. M., Sankaran, M., Higgins, S. I., Archibald, S., Hoffmann, W. A., . . . & Hutley, L. B.

(2014). Savanna vegetation-fire-climate relationships differ among continents. *Science*, 343(6170), 548–52.

Leite, M. B., Xavier, R. O., Oliveira, P. T. S., Silva, F. K. G., & Matos, D. M. S. (2018). Groundwater depth as a constraint on the woody cover in a Neotropical Savanna. *Plant and Soil*, 426(1–2), 1–15.

Lenton, T. M. (2013). Environmental tipping points. *Annual Review of Environment and Resources*, 38, 1–29.

Leonard, S. W. J., Kirkpatrick, J. B., & Marsden-Smedley, J. B. (2010). Variation between grassland structural types in the effects of vertebrate grazing on fire potential. *Journal of Applied Ecology*, 47, 876–83.

Lewontin, R. C. (1969). The meaning of stability. In *Brookhaven symposia in biology* (Vol. 22, p. 13).

Linder, H. P. (2003). The radiation of the Cape flora, southern Africa. *Biological Reviews*, 78(4), 597–638.

Linder, H. P. (2005). Evolution of diversity: the Cape flora. *Trends in Plant Science*, 10, 536–41.

Linder, H. P., Lehmann, C. E., Archibald, S., Osborne, C. P., & Richardson, D. M. (2018). Global grass (Poaceae) success underpinned by traits facilitating colonization, persistence and habitat transformation. *Biological Reviews*, 93(2), 1125–44.

Little, J. K., Prior, L. D., Williamson, G. J., Williams, S. E., & Bowman, D. M. (2012). Fire weather risk differs across rain forest—savanna boundaries in the humid tropics of north-eastern Australia. *Austral Ecology*, 37(8), 915–25.

Lloyd, J., Bird, M. I., Vellen, L., Miranda, A. C., Veenendaal, E. M., Djagbletey, G., . . . & Farquhar, G. D. (2008). Contributions of woody and herbaceous vegetation to tropical savanna ecosystem productivity: a quasi-global estimate. *Tree Physiology*, 28(3), 451–68.

Lloyd, J., Domingues, T. F., Schrodt, F., Ishida, F. Y., Feldpausch, T. R., Saiz, G., . . . & Marimon, B. S. (2015). Edaphic, structural and physiological contrasts across Amazon Basin forest–savanna ecotones suggest a role for potassium as a key modulator of tropical woody vegetation structure and function. *Biogeosciences*, 12(22), 6529–71.

Lombard, A. T., Johnson, C. F., Cowling, R. M., & Pressey, R. L. (2001). Protecting plants from elephants: botanical reserve scenarios within the Addo Elephant National Park, South Africa. *Biological Conservation*, 102(2), 191–203.

Louppe, D., Oattara, N. K., & Coulibaly, A. (1995). The effects of brush fires on vegetation: the Aubreville fire plots after 60 years. *The Commonwealth Forestry Review*, 74(4), 288–92.

Lowry, P.P., Schatz, G.E., & Phillipson, P.P. (1997) The classification of natural and anthropogenic vegetation in Madagascar. *Natural change and human impact in Madagascar* (ed. by S.M. Goodman and B.D. Patterson), pp. 93–123. Smithsonian Institution Press, Washington, DC.

Ludwig, F., De Kroon, H., Berendse, F., & Prins, H. H. (2004). The influence of savanna trees on nutrient, water and light availability and the understorey vegetation. *Plant Ecology*, 170(1), 93–105.

MacArthur, R. H., & MacArthur, J. W. (1961). On bird species diversity. *Ecology*, 42(3), 594–8.

MacFadden, B. J. (2005). Fossil horses—evidence for evolution. *Science*, 307(5716), 1728–30.

Manders, P. T. (1990). Fire and other variables as determinants of forest/fynbos boundaries in the Cape Province. *Journal of Vegetation Science*, 1(4), 483–90.

Manders, P. T., & Richardson, D. M. (1992). Colonization of Cape fynbos communities by forest species. *Forest Ecology and Management*, 48(3–4), 277–93.

Mann, C. C. (2005). *1491: New revelations of the Americas before Columbus*. Alfred A. Knopf Incorporated, New York.

Marchese, C. (2015). Biodiversity hotspots: A shortcut for a more complicated concept. *Global Ecology and Conservation*, 3, 297–309.

Maurin, O., Davies, T. J., Burrows, J. E., Daru, B. H., Yessoufou, K., Muasya, A. M., … & Bond, W. J. (2014). Savanna fire and the origins of the 'underground forests' of Africa. *New Phytologist*, 204(1), 201–14.

May, R. M. (1977). Thresholds and breakpoints in ecosystems with a multiplicity of stable states. *Nature*, 269(5628), 471.

Mayle, F. E., Beerling, D. J., Gosling, W. D., & Bush, M. B. (2004). Responses of Amazonian ecosystems to climatic and atmospheric carbon dioxide changes since the last glacial maximum. *Philosophical Transactions of the Royal Society of London Series B: Biological Sciences*, 359, 499–514.

McNaughton, S. J. (1984). Grazing lawns: animals in herds, plant form, and coevolution. *The American Naturalist*, 124(6), 863–86.

Michaletz, S. T. (2018). Xylem dysfunction in fires: towards a hydraulic theory of plant responses to multiple disturbance stressors. *New Phytologist*, 217(4), 1391–3.

Michaletz, S. T., Johnson, E. A., & Tyree, M. T. (2012). Moving beyond the cambium necrosis hypothesis of post-fire tree mortality: cavitation and deformation of xylem in forest fires. *New Phytologist*, 194(1), 254–63.

Midgley, G. F., Hannah, L., Millar, D., Rutherford, M. C., & Powrie, L. W. (2002). Assessing the vulnerability of species richness to anthropogenic climate change in a biodiversity hotspot. *Global Ecology and Biogeography*, 11(6), 445–51.

Midgley, J. J., Balfour, D., & Kerley, G. I. (2005). Why do elephants damage savanna trees?: Commentary. *South African Journal of Science*, 101(5–6), 213–15.

Midgley, J. J., Kruger, L. M., & Skelton, R. (2011). How do fires kill plants? The hydraulic death hypothesis and Cape Proteaceae 'fire-resisters'. *South African Journal of Botany*, 77(2), 381–6.

Midgley, J. J., & Niklas, K. J. (2004). Does disturbance prevent total basal area and biomass in indigenous forests from being at equilibrium with the local environment? *Journal of Tropical Ecology*, 20(5), 595–7.

Miehe, G., Schleuss, P. M., Seeber, E., Babel, W., Biermann, T., Braendle, M., … & Graf, H. F. (2019). The *Kobresia pygmaea* ecosystem of the Tibetan Highlands: origin, functioning, and degradation of the world's largest pastoral alpine ecosystem. *Science of the Total Environment*, 648, 754–71.

Milchunas, D. G., & Lauenroth, W. K. (1993). Quantitative effects of grazing on vegetation and soils over a global range of environments: ecological archives M063–001. *Ecological Monographs*, 63(4), 327–66.

Mills, A. J., Milewski, A. V., & Sirami, C. (2016). A preliminary test of catabolic nutrients in explanation of the puzzling treelessness of grassland in mesic Australia. *Austral Ecology*, 41(8), 927–37.

Mills, A. J., Milewski, A. V., Snyman, D., & Jordaan, J. J. (2017). Effects of anabolic and catabolic nutrients on woody plant encroachment after long-term experimental fertilization in a South African savanna. *PloS One*, 12(6), e0179848.

Mitchell, F. J. (2005). How open were European primeval forests? Hypothesis testing using palaeoecological data. *Journal of Ecology*, 93(1), 168–77.

Mittermeier, R. A., Turner, W. R., Larsen, F. W., Brooks, T. M., & Gascon, C. (2011). Global biodiversity conservation: the critical role of hotspots. In FE Zachos & JC Habel; *Biodiversity hotspots* (pp. 3–22). Springer, Berlin, Heidelberg. doi: 10.1007/978-3-642-20992-5_1

Moncrieff, G. R., Chamaillé-Jammes, S., & Bond, W. J. (2014a). Modelling direct and indirect impacts of browser consumption on woody plant growth: moving beyond biomass. *Oikos*, 123(3), 315–22.

Moncrieff, G. R., Kruger, L. M., & Midgley, J. J. (2008). Stem mortality of *Acacia nigrescens* induced by the synergistic effects of elephants and fire in Kruger National Park, South Africa. *Journal of Tropical Ecology*, 24(6), 655–62.

Moncrieff, G. R., Lehmann, C. E., Schnitzler, J., Gambiza, J., Hiernaux, P., Ryan, C. M., Shackleton, C. M., Williams, R. J., & Higgins, S. I. (2014c). Contrasting architecture of key African and Australian savanna tree taxa drives intercontinental structural divergence. *Global Ecology and Biogeography*, 23(11), 1235–44.

Moncrieff, G. R., Scheiter, S., Bond, W. J., & Higgins, S. I. (2014b). Increasing atmospheric CO_2 overrides the historical legacy of multiple stable biome states in Africa. *New Phytologist*, 201(3), 908–15.

Moncrieff, G. R., Scheiter, S., Langan, L., Trabucco, A., & Higgins, S. I. (2016). The future distribution of the savannah biome: model-based and biogeographic contingency. *Philosophical Transactions of the Royal Society of London B: Biological Sciences*, 371(1703), 20150311.

Moolman, H. J., & Cowling, R. M. (1994). The impact of elephant and goat grazing on the endemic flora of South African succulent thicket. *Biological Conservation*, 68(1), 53–61.

Mooney, H. A. (1977). Frost sensitivity and resprouting behavior of analogous shrubs of California and Chile. *Madroño*, 24(2), 74–8.

Morgan, J. A., Pataki, D. E., Körner, C., Clark, H., Del Grosso, S. J., Grünzweig, J. M.,…& Nippert, J. B. (2004). Water relations in grassland and desert ecosystems exposed to elevated atmospheric CO_2. *Oecologia*, 140(1), 11–25.

Morley, R. J., & Richards, K. (1993). Gramineae cuticle: a key indicator of Late Cenozoic climatic change in the Niger Delta. *Review of Palaeobotany and Palynology*, 77(1–2), 119–27.

Mucina, L., & Wardell-Johnson, G. W. (2011). Landscape age and soil fertility, climatic stability, and fire regime predictability: beyond the OCBIL framework. *Plant and Soil*, 341(1–2), 1–23.

Murphy, B. P., Andersen, A. N., & Parr, C. L. (2016). The underestimated biodiversity of tropical grassy biomes. *Philosophical Transactions of the Royal Society of London B: Biological Sciences*, 371(1703), 20150319.

Murphy, B.P., and Bowman, D.M. (2012). What controls the distribution of tropical forest and savanna?. *Ecology Letters*, 15(7), 748–58.

Mutch, R. W. (1970). Wildland fires and ecosystems: a hypothesis. *Ecology*, 51(6), 1046–51.

Myers, N., Mittermeier, R. A., Mittermeier, C. G., Da Fonseca, G. A., & Kent, J. (2000). Biodiversity hotspots for conservation priorities. *Nature*, 403(6772), 853.

Nepstad, D. C., Verssimo, A., Alencar, A., Nobre, C., Lima, E., Lefebvre, P., Schlesinger, P., Potter, C., Moutinho, P., Mendoza, E., & Cochrane, M. (1999). Large-scale impoverishment of Amazonian forests by logging and fire. *Nature*, 398(6727), 505.

Nihlgård, B. (1971). Pedological influence of spruce planted on former beech forest soils in Scania, South Sweden. *Oikos*, 22(3), 302–14.

Nogueira, C., Ribeiro, S., Costa, G. C., & Colli, G. R. (2011). Vicariance and endemism in a Neotropical savanna hotspot: distribution patterns of Cerrado squamate reptiles. *Journal of Biogeography*, 38(10), 1907–22.

Noss, R. F. (2012). *Forgotten grasslands of the South: natural history and conservation*. Island Press, Washington, DC.

Noss, R. F., Platt, W. J., Sorrie, B. A., Weakley, A. S., Means, D. B., Costanza, J., & Peet, R. K. (2015). How global biodiversity hotspots may go unrecognized: lessons from the North American Coastal Plain. *Diversity and Distributions*, 21(2), 236–44.

Nowacki, G. J., & Abrams, M. D. (2008). The demise of fire and 'mesophication' of forests in the eastern United States. *AIBS Bulletin*, 58(2), 123–38.

Nykvist, N. (2000). Tropical forests can suffer from a serious deficiency of calcium after logging. *Ambio: A Journal of the Human Environment*, 29(6), 310–13.

O'Connor, T. G., Puttick, J. R., & Hoffman, M. T. (2014). Bush encroachment in southern Africa: changes and causes. *African Journal of Range & Forage Science*, 31(2), 67–88.

O'Kane, C. A., Duffy, K. J., Page, B. R., & Macdonald, D. W. (2012). Heavy impact on seedlings by the impala suggests a central role in woodland dynamics. *Journal of Tropical Ecology*, 28(3), 291–7.

Odion, D. C., Moritz, M. A., & DellaSala, D. A. (2010). Alternative community states maintained by fire in the Klamath Mountains, USA. *Journal of Ecology*, 98(1), 96–105.

Odling-Smee, F. J., Laland, K. N., & Feldman, M. W. (1996). Niche construction. *American Naturalist*, 147(4), 641–8.

Odling-Smee, F. J., Laland, K. N., & Feldman, M. W. (2003). *Niche construction: the neglected process in evolution*. Monographs in population biology (ISSN 0077–0930, (37). Princeton University Press, Princeton, NJ.

Olson, D. M., Dinerstein, E., Wikramanayake, E. D., Burgess, N. D., Powell, G. V., Underwood, E. C.,…& Loucks, C. J. (2001). Terrestrial Ecoregions of the World: A New Map of Life on Earth. A new global map of terrestrial ecoregions provides an innovative tool for conserving biodiversity. *BioScience*, 51(11), 933–8.

Ondei, S., Prior, L. D., Vigilante, T., & Bowman, D. M. (2017). Fire and cattle disturbance affects vegetation structure and rain forest expansion into savanna in the Australian monsoon tropics. *Journal of Biogeography*, 44(10), 2331–42.

Orians, G. H., & Milewski, A. V. (2007). Ecology of Australia: the effects of nutrient-poor soils and intense fires. *Biological Reviews*, 82(3), 393–423.

Osborne, C. P. (2008). Atmosphere, ecology and evolution: what drove the Miocene expansion of C4 grasslands? *Journal of Ecology*, 96, 35–45.

Overbeck, G. E., Müller, S. C., Fidelis, A., Pfadenhauer, J., Pillar, V. D., Blanco, C. C.,…& Forneck, E. D. (2007). Brazil's neglected biome: the South Brazilian Campos. *Perspectives in Plant Ecology, Evolution and Systematics*, 9(2), 101–16.

Ovington, J.D. (1953). Studies of the development of woodland conditions under different trees. Part I. Soil pH. *Journal of Ecology*, 41, 13–34.

Owen-Smith, N. (1987). Pleistocene extinctions: the pivotal role of megaherbivores. *Paleobiology*, 13, 351–62.

Owen-Smith, R. N. (1988). *Megaherbivores: the influence of very large body size on ecology*. Cambridge University Press, Cambridge, UK.

Owen-Smith, N. (2013). Contrasts in the large herbivore faunas of the southern continents in the late Pleistocene and the ecological implications for human origins. *Journal of Biogeography*, 40(7), 1215–24.

Pachzelt, A., Forrest, M., Rammig, A., Higgins, S. I., & Hickler, T. (2015). Potential impact of large ungulate grazers on African vegetation, carbon storage, and fire regimes. *Global Ecology and Biogeography*, 24(9), 991–1002.

Palmer, S. C. F., & Truscott, A. M. (2003). Browsing by deer on naturally regenerating Scots pine (*Pinus sylvestris* L.) and its effects on sapling growth. *Forest Ecology and Management*, 182(1–3), 31–47.

Paoli, G. D., Curran, L. M., & Slik, J. W. F. (2008). Soil nutrients affect spatial patterns of aboveground biomass and emergent tree density in southwestern Borneo. *Oecologia*, 155(2), 287–99.

Papanastasis, V. P., Arianoutsou, M, & Papanastasis, K. (2010). Environmental conservation in classical Greece. *Journal of Biological Research–Thessaloniki*, 14, 123–35.

Parr, C. L., Gray, E. F., & Bond, W. J. (2012). Cascading biodiversity and functional consequences of a global change–induced biome switch. *Diversity and Distributions*, 18(5), 493–503.

Parr, C. L., Lehmann, C. E., Bond, W. J., Hoffmann, W. A., & Andersen, A. N. (2014). Tropical grassy biomes: misunderstood, neglected, and under threat. *Trends in Ecology & Evolution*, 29(4), 205–13.

Pausas, J. G. (2015). Bark thickness and fire regime. *Functional Ecology*, 29(3), 315–27.

Pausas, J. G., Alessio, G. A., Moreira, B., & Corcobado, G. (2012). Fires enhance flammability in *Ulex parviflorus*. *New Phytologist*, 193(1), 18–23.

Pausas, J. G., Bradstock, R. A., Keith, D. A., & Keeley, J. E. (2004). Plant functional traits in relation to fire in crown-fire ecosystems. *Ecology*, 85(4), 1085–100.

Pausas, J. G., & Fernández-Muñoz, S. (2012). Fire regime changes in the Western Mediterranean Basin: from fuel-limited to drought-driven fire regime. *Climatic Change*, 110(1–2), 215–26.

Pausas, J. G., & Keeley, J. E. (2009). A burning story: the role of fire in the history of life. *BioScience*, 59(7), 593–601.

Pausas, J. G., & Keeley, J. E. (2017). Epicormic resprouting in fire-prone ecosystems. *Trends in Plant Science*, 22(12), 1008–15.

Pausas, J. G., Keeley, J. E., & Schwilk, D. W. (2017). Flammability as an ecological and evolutionary driver. *Journal of Ecology*, 105(2), 289–97.

Peck, A. J. (1978). Salinization of non-irrigated soils and associated streams: a review. *Soil Research*, 16(2), 157–68.

Perrier de la Bâthie, H. (1928) Les prairies de Madagascar. *Revue de Botanique Appliquée & d'Agriculture Coloniale* (in 3 parts), 8, 549–57; 8, 631–42; 8, 696–707.

Peterken, G. (2009). Response to 'A postulated natural origin for the open landscape of upland Scotland'. *Plant Ecology & Diversity*, 2(1), 89–90.

Peterson, D. W., & Reich, P. B. (2001). Prescribed fire in oak savanna: fire frequency effects on stand structure and dynamics. *Ecological Applications*, 11(3), 914–27.

Peterson, D. W., & Reich, P. B. (2008). Fire frequency and tree canopy structure influence plant species diversity in a forest-grassland ecotone. *Plant Ecology*, 194(1), 5–16.

Petraitis, P. (2013). *Multiple stable states in natural ecosystems*. Oxford University Press, Oxford, UK.

Phillips, J. (1936). Fire in vegetation: a bad master, a good servant, and a national problem. *S. Afr. J. Bot.*, 2, 36–45.

Phillips, J. F. V. (1930). Fire: its influence on biotic communities and physical factors in South and East Africa. *South African Journal of Science*, 27, 352–67.

Pinheiro, E. D. S., & Durigan, G. (2009). Dinâmica espaço-temporal (1962–2006) das fitofisionomias em unidade de conservação do Cerrado no sudeste do Brasil. *Revista Brasileira de Botânica*, 32(3), 441–54.

Polis, G. A. (1999). Why are parts of the world green? Multiple factors control productivity and the distribution of biomass. *Oikos*, 86, 3–15.

Polley, H. W., Mayeux, H. S., Johnson, H. B., & Tischler, C. R. (1997). Atmospheric CO_2, soil water, and shrub/grass ratios on rangelands. *Journal of Range Management*, 50, 278–84.

Poulsen, Z. C., & Hoffman, M. T. (2015). Changes in the distribution of indigenous forest in Table Mountain National Park during the 20th Century. *South African Journal of Botany*, 101, 49–56.

Prentice, I. C., & Harrison, S. P. (2009). Ecosystem effects of CO_2 concentration: evidence from past climates. *Climate of the Past*, 5(3), 297–307.

Pringle, H., Zimmerman, I., & Tinley, K. (2011). Accelerating landscape incision and the downward spiralling rain use efficiency of Namibian rangelands. *Agricola*, 21, 43–52.

Prins, H. H., & van der Jeugd, H. P. (1993). Herbivore population crashes and woodland structure in East Africa. *Journal of Ecology*, 81, 305–14.

Pyne, S. J. (1990). Fire conservancy: The origins of wildland fire protection in British India, America, and Australia. In *Fire in the Tropical Biota* (ed. J.G. Goldammer), Ecological Studies 84, (pp. 319–36). Springer, Berlin.

Quéméré, E., Amelot, X., Pierson, J., Crouau-Roy, B., & Chikhi, L. (2012). Genetic data suggest a natural prehuman origin of open habitats in northern Madagascar and question the deforestation narrative in this region. *Proceedings of the National Academy of Sciences of the United States of America*, 109(32), 13028–33.

Quesada, C. A., Lloyd, J., Schwarz, M., Baker, T. R., Phillips, O. L., et al. (2009). Regional and large-scale patterns in Amazon forest structure and function are mediated by variations in soil physical and chemical properties. *Biogeosciences Discussions*, 6, 3993–4057.

Quirk, J., McDowell, N. G., Leake, J. R., Hudson, P. J., & Beerling, D. J. (2013). Increased susceptibility to drought-induced mortality in *Sequoia sempervirens* (Cupressaceae) trees under Cenozoic atmospheric carbon dioxide starvation. *American Journal of Botany*, 100(3), 582–91.

Ratnam, J., Bond, W. J., Fensham, R. J., Hoffmann, W. A., Archibald, S., Lehmann, C. E., Anderson, M.T., Higgins,

S.I., and Sankaran, M. (2011). When is a 'forest' a savanna, and why does it matter? *Global Ecology and Biogeography*, 20(5), 653–60.

Ratnam, J., Tomlinson, K. W., Rasquinha, D. N., & Sankaran, M. (2016). Savannahs of Asia: antiquity, biogeography, and an uncertain future. *Philosophical Transactions of the Royal Society of London B: Biological Sciences*, 371(1703), 20150305.

Ratter, J. A., Ribeiro, J. F., & Bridgewater, S. (1997). The Brazilian cerrado vegetation and threats to its biodiversity. *Annals of Botany*, 80, 223–30.

Richardson, D. M., & Rejmánek, M. (2011). Trees and shrubs as invasive alien species–a global review. *Diversity and Distributions*, 17(5), 788–809.

Richardson, D. M., & Van Wilgen, B. W. (1986). Effects of thirty-five years of afforestation with *Pinus radiata* on the composition of mesic mountain fynbos near Stellenbosch. *South African Journal of Botany*, 52(4), 309–15.

Riginos, C. (2009). Grass competition suppresses savanna tree growth across multiple demographic stages. *Ecology*, 90(2), 335–40.

Ripley, B., Visser, V., Christin, P. A., Archibald, S., Martin, T., & Osborne, C. (2015). Fire ecology of C3 and C4 grasses depends on evolutionary history and frequency of burning but not photosynthetic type. *Ecology*, 96(10), 2679–91.

Ripple, W. J., & Beschta, R. L. (2004). Wolves and the ecology of fear: can predation risk structure ecosystems? *BioScience*, 54(8), 755–66.

Ripple, W. J., & Beschta, R. L. (2012). Trophic cascades in Yellowstone: the first 15 years after wolf reintroduction. *Biological Conservation*, 145(1), 205–13.

Robinson, G. S., Pigott Burney, L., & Burney, D. A. (2005). Landscape paleoecology and megafaunal extinction in southeastern New York State. *Ecological Monographs*, 75(3), 295–315.

Rogers, B. M., Soja, A. J., Goulden, M. L., & Randerson, J. T. (2015). Influence of tree species on continental differences in boreal fires and climate feedbacks. *Nature Geoscience*, 8(3), 228.

Rosell, J. A. (2016). Bark thickness across the angiosperms: more than just fire. *New Phytologist*, 211(1), 90–102.

Ruggiero, P. G. C., Batalha, M. A., Pivello, V. R., & Meirelles, S. T. (2002). Soil-vegetation relationships in cerrado (Brazilian savanna) and semideciduous forest, Southeastern Brazil. *Plant Ecology*, 160(1), 1–16.

Rule, S., Brook, B. W., Haberle, S. G., Turney, C. S., Kershaw, A. P., & Johnson, C. N. (2012). The aftermath of megafaunal extinction: ecosystem transformation in Pleistocene Australia. *Science*, 335(6075), 1483–6.

Rundel, P. W., Arroyo, M. T., Cowling, R. M., Keeley, J. E., Lamont, B. B., Pausas, J. G., & Vargas, P. (2018). Fire and plant diversification in Mediterranean-climate regions.

Frontiers in plant science, 9. https://doi.org/10.3389/fpls .2018.00851

Rundel, P. W., Arroyo, M. T., Cowling, R. M., Keeley, J. E., Lamont, B. B., & Vargas, P. (2016). Mediterranean biomes: evolution of their vegetation, floras, and climate. *Annual Review of Ecology, Evolution, and Systematics*, 47, 383–407.

Saarinen, J. J., Boyer, A. G., Brown, J. H., Costa, D. P., Ernest, S. M., Evans, A. R.,…& Lintulaakso, K. (2014). Patterns of maximum body size evolution in Cenozoic land mammals: eco-evolutionary processes and abiotic forcing. *Proceedings of the Royal Society of London B: Biological Sciences*, 281(1784), 20132049.

Sage, R. F., & Kubien, D. S. (2003). Quo vadis C4? An eco-physiological perspective on global change and the future of C4 plants. *Photosynthesis Research*, 77(2–3), 209–25.

San José, J. J., & Fariñas, M. R. (1983). Changes in tree density and species composition in a protected *Trachypogon savanna*, Venezuela. *Ecology*, 64(3), 447–53.

Sanders, W.J., Gheerbrant, E., Harris, J. M., Saegusa, H., & Delmer, C. (2010). Proboscidea. In: Werdelin, L., & Sanders, W. J. (2010). *Cenozoic mammals of Africa*. University of California Press, Berkeley. Pp. 161–251.

Sankaran, M., Augustine, D. J., & Ratnam, J. (2013). Native ungulates of diverse body sizes collectively regulate long-term woody plant demography and structure of a semi-arid savanna. *Journal of Ecology*, 101(6), 1389–99.

Sankaran, M., Hanan, N. P., Scholes, R. J., Ratnam, J., Augustine, D. J., Cade, B. S., Gignoux, J., Higgins, S. I., Le Roux, X., Ludwig, F., et al. (2005). Determinants of woody cover in African savannas. *Nature*, 438: 846–9.

Sankaran, M., Ratnam, J., & Hanan, N. P. (2004). Tree–grass coexistence in savannas revisited–insights from an examination of assumptions and mechanisms invoked in existing models. *Ecology Letters*, 7(6), 480–90.

Sankaran, M., Ratnam, J., & Hanan, N. (2008). Woody cover in African savannas: the role of resources, fire and herbivory. *Global Ecology and Biogeography*, 17(2), 236–45.

San-Miguel-Ayanz, J., Moreno, J. M., & Camia, A. (2013). Analysis of large fires in European Mediterranean landscapes: lessons learned and perspectives. *Forest Ecology and Management*, 294, 11–22.

Sarmiento G. (1984). *The ecology of Neotropical savannas*. Harvard University Press, Cambridge, MA.

Sauer, C. O. (1950). Grassland climax, fire, and man. *Journal of Range Management*, 3(1), 16–21.

Savory, A., & Butterfield, J. (1998). *Holistic management: a new framework for decision making*. Island Press, Washington, DC.

Scheffer, M., & Carpenter, S. R. (2003). Catastrophic regime shifts in ecosystems: linking theory to observation. *Trends in Ecology & Evolution*, 18(12), 648–56.

Scheffer, M., Carpenter, S.R., Lenton, T.M., Bascompte, J., Brock, W., Dakos, V., van de Koppel, J., van de Leemput,

I.A., Levin, S.A., van Nes, E.H., & Pascual, M. (2012a). Anticipating critical transitions. *Science*, 338(6105), 344–8.

Scheffer, M., Hirota, M., Holmgren, M., Van Nes, E. H., & Chapin, F. S. (2012b). Thresholds for boreal biome transitions. *Proceedings of the National Academy of Sciences of the United States of America*, 109(52), 21384–9.

Scheiter, S., and Higgins, S.I. (2009). Impacts of climate change on the vegetation of Africa: an adaptive dynamic vegetation modelling approach. *Global Change Biology*, 15(9), 2224–46.

Scheiter, S., Higgins, S. I., Osborne, C. P., Bradshaw, C., Lunt, D., Ripley, B. S., Taylor L.L., & Beerling, D. J. (2012). Fire and fire-adapted vegetation promoted C4 expansion in the late Miocene. *New Phytologist*, 195(3), 653–66.

Scheiter, S., Langan, L., & Higgins, S. I. (2013). Next-generation dynamic global vegetation models: learning from community ecology. *New Phytologist*, 198(3), 957–69.

Schimper, A.F.W. (1903). *Plant geography on a physiological basis*. Oxford.

Scholes, R. J. (1990). The influence of soil fertility on the ecology of southern African dry savannas. *Journal of Biogeography*, 17(4/5), 415–19.

Scholes, R. J., & Archer, S. R. (1997). Tree-grass interactions in savannas. *Annual Review of Ecology and Systematics*, 28(1), 517–44.

Schonenberger, J. (2005). Rise from the ashes: the reconstruction of charcoal fossil flowers. *Trends in Plant Science*, 10, 436–43.

Schutz, A. E. N., Bond, W. J., & Cramer, M. D. (2009). Juggling carbon: allocation patterns of a dominant tree in a fire-prone savanna. *Oecologia*, 160(2), 235.

Schwilk, D. W. (2015). Dimensions of plant flammability. *New Phytologist*, 206(2), 486–8.

Schwilk, D. W., & Ackerly, D. D. (2001). Flammability and serotiny as strategies: correlated evolution in pines. *Oikos*, 94(2), 326–36.

Schwilk, D. W., & Caprio, A. C. (2011). Scaling from leaf traits to fire behaviour: community composition predicts fire severity in a temperate forest. *Journal of Ecology*, 99(4), 970–80.

Scott, A. C. (2010). Charcoal recognition, taphonomy and uses in palaeoenvironmental analysis. *Palaeogeography, Palaeoclimatology, Palaeoecology*, 291(1–2), 11–39.

Scott, A. C. (2018). *Burning planet: the story of fire through time*. Oxford University Press, Oxford, UK.

Scott, A. C., Bowman, D. M., Bond, W. J., Pyne, S. J., & Alexander, M. E. (2014). *Fire on Earth: an introduction*. Wiley-Blackwell, Oxford, UK.

Shugart, H. H., & Woodward, F. I. (2010). *Global change and the terrestrial biosphere: achievements and challenges*. John Wiley & Sons.

Silva, L. C., Hoffmann, W. A., Rossatto, D. R., Haridasan, M., Franco, A. C., & Horwath, W. R. (2013). Can savannas become forests? A coupled analysis of nutrient stocks and fire thresholds in central Brazil. *Plant and Soil*, 373(1–2), 829–42.

Silva, L. C., Sternberg, L., Haridasan, M., Hoffmann, W. A., Miralles-Wilhem, F., & Franco, A. C. (2008). Expansion of gallery forests into central Brazilian savannas. *Global Change Biology*, 14(9), pp. 2108–18.

Silveira, F. A., Negreiros, D., Barbosa, N. P., Buisson, E., Carmo, F. F., Carstensen, D. W.,...& Garcia, Q. S. (2016). Ecology and evolution of plant diversity in the endangered campo rupestre. *Plant and Soil*, 403(1–2), 129–52.

Simon, M. F., Grether, R., de Queiroz, L. P., Skema, C., Pennington, R. T., & Hughes, C. E. (2009). Recent assembly of the Cerrado, a neotropical plant diversity hotspot, by *in situ* evolution of adaptations to fire. *Proceedings of the National Academy of Sciences of the United States of America*, 106(48), 20359–64.

Simon, M. F., & Pennington, T. (2012). Evidence for adaptation to fire regimes in the tropical savannas of the Brazilian Cerrado. *International Journal of Plant Sciences*, 173(6), 711–23.

Simpson, K. J., Ripley, B. S., Christin, P. A., Belcher, C. M., Lehmann, C. E., Thomas, G. H., & Osborne, C. P. (2016). Determinants of flammability in savanna grass species. *Journal of Ecology*, 104(1), 138–48.

Sinclair, A. R. E. (1979). The eruption of the ruminants. In Sinclair, A. R. E., Norton-Griffiths, M. (eds), *Serengeti: dynamics of an ecosystem*. Chicago: Chicago University Press. pp. 82–103.

Sinclair, A. R. E., Mduma, S., & Brashares, J. S. (2003). Patterns of predation in a diverse predator–prey system. *Nature*, 425(6955), 288.

Sirami, C., Seymour, C., Midgley, G., & Barnard, P. (2009). The impact of shrub encroachment on savanna bird diversity from local to regional scale. *Diversity and Distributions*, 15(6), 948–57.

Skinner, J. D. (1993). Springbok (*Antidorcas marsupialis*) treks. *Transactions of the Royal Society of South Africa*, 48(2), 291–305.

Smit, I. P., Asner, G. P., Govender, N., Kennedy-Bowdoin, T., Knapp, D. E., & Jacobson, J. (2010). Effects of fire on woody vegetation structure in African savanna. *Ecological Applications*, 20(7), 1865–75.

Smit, I. P., Asner, G. P., Govender, N., Vaughn, N. R., & Wilgen, B. W. (2016). An examination of the potential efficacy of high-intensity fires for reversing woody encroachment in savannas. *Journal of Applied Ecology*, 53(5), 1623–33.

Smit, I. P., & Prins, H. H. (2015). Predicting the effects of woody encroachment on mammal communities, grazing biomass and fire frequency in African savannas. *PloS One*, 10(9), e0137857.

Soderberg, K., & Compton, J. S. (2007). Dust as a nutrient source for fynbos ecosystems, South Africa. *Ecosystems*, 10(4), 550–61.

Specht, R. L. (1979) Heathlands and related shrublands of the world. In: Specht, R. L. (ed.), *Ecosystems of the world, vol 9A. Heathlands and related shrublands: descriptive studies*. Elsevier, Amsterdam, pp. 1–18.

Specht, R. L., & Moll, E. J. (1983). Mediterranean-type heathlands and sclerophyllous shrublands of the world: an overview. In: F.J. Kruger (ed.), *Mediterranean-type Ecosystems* (pp. 41–65). Springer, Berlin, Heidelberg.

Staver, A. C., Archibald, S., & Levin, S. A. (2011a). The global extent and determinants of savanna and forest as alternative biome states. *Science*, 334(6053), 230–2.

Staver, A. C., Archibald, S., & Levin, S. (2011b). Tree cover in sub-Saharan Africa: rainfall and fire constrain forest and savanna as alternative stable states. *Ecology*, 92(5), 1063–72.

Staver, A. C., & Bond, W. J. (2014). Is there a 'browse trap'? Dynamics of herbivore impacts on trees and grasses in an African savanna. *Journal of Ecology*, 102(3), 595–602.

Staver, A. C., Bond, W. J., Cramer, M. D., & Wakeling, J. L. (2012). Top-down determinants of niche structure and adaptation among African Acacias. *Ecology Letters*, 15(7), 673–9.

Staver, A. C., & Hansen, M. C. (2015). Analysis of stable states in global savannas: is the CART pulling the horse?– a comment. *Global Ecology and Biogeography*, 24(8), 985–7.

Stevens, N., Erasmus, B. F. N., Archibald, S., & Bond, W. J. (2016). Woody encroachment over 70 years in South African savannahs: overgrazing, global change or extinction aftershock? *Philosophical Transactions of the Royal Society of London B: Biological Sciences*, 371(1703), 20150437.

Stevens, N., Lehmann, C. E., Murphy, B. P., & Durigan, G. (2017). Savanna woody encroachment is widespread across three continents. *Global Change Biology*, 23(1), 235–44.

Sukumar, R. (1990). Ecology of the Asian elephant in southern India. II. Feeding habits and crop raiding patterns. *Journal of Tropical Ecology*, 6(1), 33–53.

Svenning, J. C. (2002). A review of natural vegetation openness in north-western Europe. *Biological Conservation*, 104(2), 133–48.

Svenning, J. C., Pedersen, P. B., Donlan, C. J., Ejrnæs, R., Faurby, S., Galetti, M., ... & Vera, F. W. (2016). Science for a wilder Anthropocene: Synthesis and future directions for trophic rewilding research. *Proceedings of the National Academy of Sciences of the United States of America*, 113(4), 898–906.

Takatsuki, S. (2009). Effects of sika deer on vegetation in Japan: a review. *Biological Conservation*, 142(9), 1922–9.

Takyu, M., Aiba, S. I., & Kitayama, K. (2003). Changes in biomass, productivity and decomposition along topographical gradients under different geological conditions in tropical lower montane forests on Mount Kinabalu, Borneo. *Oecologia*, 134(3), 397–404.

Tanentzap, A. J., Bazely, D. R., Koh, S., Timciska, M., Haggith, E. G., Carleton, T. J., & Coomes, D. A. (2011). Seeing the forest for the deer: do reductions in deer-disturbance lead to forest recovery? *Biological Conservation*, 144(1), 376–82.

Tanentzap, A. J., & Coomes, D. A. (2012). Carbon storage in terrestrial ecosystems: do browsing and grazing herbivores matter? *Biological Reviews*, 87(1), 72–94.

Tanentzap, A. J., Zou, J., & Coomes, D. A. (2013). Getting the biggest birch for the bang: restoring and expanding upland birchwoods in the Scottish Highlands by managing red deer. *Ecology and Evolution*, 3(7), 1890–901.

Tang, Z. H., & Ding, Z. L. (2013). A palynological insight into the Miocene aridification in the Eurasian interior. *Palaeoworld*, 22(3–4), 77–85.

Terborgh, J., Davenport, L. C., Niangadouma, R., Dimoto, E., Mouandza, J. C., Scholtz, O., & Jaen, M. R. (2016a). Megafaunal influences on tree recruitment in African equatorial forests. *Ecography*, 39(2), 180–6.

Terborgh, J., Davenport, L. C., Niangadouma, R., Dimoto, E., Mouandza, J. C., Schultz, O., & Jaen, M. R. (2016b). The African rainforest: odd man out or megafaunal landscape? African and Amazonian forests compared. *Ecography*, 39(2), 187–93.

Terborgh, J., Davenport, L. C., Ong, L., & Campos-Arceiz, A. (2018). Foraging impacts of Asian megafauna on tropical rain forest structure and biodiversity. *Biotropica*, 50(1), 84–9.

Thuiller, W., Albert, C., Araújo, M. B., Berry, P. M., Cabeza, M., Guisan, A., Hickler, T., Midgley GF, Paterson J, Schurr FM, Sykes MT, & Zimmermann, N. E. (2008). Predicting global change impacts on plant species' distributions: future challenges. *Perspectives in Plant Ecology, Evolution and Systematics*, 9(3), 137–52.

Thuiller, W., Lavorel, S., Araújo, M. B., Sykes, M. T., & Prentice, I. C. (2005). Climate change threats to plant diversity in Europe. *Proceedings of the National Academy of Sciences of the United States of America*, 102(23), 8245–50.

Tiffney, B. H. (1984). Seed size, dispersal syndromes, and the rise of the angiosperms: evidence and hypothesis. *Annals of the Missouri Botanical Garden*, 71(2), 551–76.

Tilman, D. (1982). *Resource competition and community structure*. Princeton University Press, Princeton, NJ.

Tilman, D., Reich, P. B., & Knops, J. M. (2006). Biodiversity and ecosystem stability in a decade-long grassland experiment. *Nature*, 441(7093), 629.

Tilman, D., & Wedin, D. (1991). Oscillations and chaos in the dynamics of a perennial grass. *Nature*, 353(6345), 653.

Tng, D. Y. P., Williamson, G. J., Jordan, G. J., & Bowman, D. M. J. S. (2012). Giant eucalypts—globally unique fire-adapted rain-forest trees? *New Phytologist*, 196(4), 1001–14.

Tunison, J. T., D'Antonio, C. M., & Loh, R. K. (2000). Fire and invasive plants in Hawai'i Volcanoes National Park. In *Proceedings of the invasive species workshop: the role of fire in the control and spread of invasive species. Fire conference* (pp. 122–31).

Turpie, J. K., Marais, C., & Blignaut, J. N. (2008). The working for water programme: Evolution of a payments for ecosystem services mechanism that addresses both poverty and ecosystem service delivery in South Africa. *Ecological Economics*, 65(4), 788–98.

Twidwell, D., Rogers, W. E., Fuhlendorf, S. D., Wonkka, C. L., Engle, D. M., Weir, J. R., Kreuter, U.P., & Taylor, Jr, C. A. (2013). The rising Great Plains fire campaign: citizens' response to woody plant encroachment. *Frontiers in Ecology and the Environment*, 11(s1), e64–e71.

Valeix, M., Loveridge, A. J., Chamaillé-Jammes, S., Davidson, Z., Murindagomo, F., Fritz, H., & Macdonald, D. W. (2009). Behavioral adjustments of African herbivores to predation risk by lions: spatiotemporal variations influence habitat use. *Ecology*, 90(1), 23–30.

Van Breemen, N. (1995). How Sphagnum bogs down other plants. *Trends in Ecology & Evolution*, 10(7), 270–5.

van der Plas, F., Howison, R. A., Mpanza, N., Cromsigt, J. P., & Olff, H. (2016). Different-sized grazers have distinctive effects on plant functional composition of an African savannah. *Journal of Ecology*, 104(3), 864–75.

Van der Putten, W. H., Bardgett, R. D., Bever, J. D., Bezemer, T. M., Casper, B. B., Fukami, T., Kardol P., Klironomos, J.N., Kulmatiski, A., Schweitzer, J.A., Suding, K. N., van der Voorde, T.F.J., & Wardle, D.A. (2013). Plant–soil feedbacks: the past, the present and future challenges. *Journal of Ecology*, 101(2), 265–76.

Van der Werf, G. R., Randerson, J. T., Giglio, L., Collatz, G. J., Mu, M., Kasibhatla, P. S., . . . & van Leeuwen, T. T. (2010). Global fire emissions and the contribution of deforestation, savanna, forest, agricultural, and peat fires (1997–2009). *Atmospheric Chemistry and Physics*, 10(23), 11707–35.

Van Nes, E. H., Hirota, M., Holmgren, M., & Scheffer, M. (2014). Tipping points in tropical tree cover: linking theory to data. *Global Change Biology*, 20(3), 1016–21.

Van Wilgen, B. W., Higgins, K. B., & Bellstedt, D. U. (1990). The role of vegetation structure and fuel chemistry in excluding fire from forest patches in the fire-prone fynbos shrublands of South Africa. *Journal of Ecology*, 78(1), 210–22.

Van Wyk, D. B. (1987). Some effects of afforestation on streamflow in the Western Cape Province, South Africa. *Water SA*, 13(1), 31–6.

Vandenberghe, C., Freléchoux, F., Gadallah, F., & Buttler, A. (2006). Competitive effects of herbaceous vegetation on tree seedling emergence, growth and survival: Does gap size matter? *Journal of Vegetation Science*, 17(4), 481–8.

Vasconcelos, H. L., Maravalhas, J. B., & Cornelissen, T. (2017). Effects of fire disturbance on ant abundance and diversity: A global meta-analysis. *Biodiversity and Conservation*, 26, 177–88.

Veblen, K. E., Porensky, L. M., Riginos, C., & Young, T. P. (2016). Are cattle surrogate wildlife? Savanna plant community composition explained by total herbivory more than herbivore type. *Ecological Applications*, 26(6), 1610–23.

Veenendaal, E. M., Torello-Raventos, M., Miranda, H. S., Sato, N. M., Oliveras, I., van Langevelde, F., . . . & Lloyd, J. (2018). On the relationship between fire regime and vegetation structure in the tropics. *New Phytologist*, 218(1), 153–66.

Veldman, J. W. (2016). Clarifying the confusion: old-growth savannahs and tropical ecosystem degradation. *Philosophical Transactions of the Royal Society of London B: Biological Sciences*, 371(1703), 20150306.

Veldman, J. W., Buisson, E., Durigan, G., Fernandes, G. W., Le Stradic, S., Mahy, G., . . . Bond, W. J. (2015a). Toward an old-growth concept for grasslands, savannas, and woodlands. *Frontiers in Ecology and the Environment*, 13, 146–53.

Veldman, J. W., Overbeck, G. E., Negreiros, D., Mahy, G., Le Stradic, S., Fernandes, G. W., Durigan, G., Buisson, E., Putz, F. E., & Bond, W.J. (2015b). Where tree planting and forest expansion are bad for biodiversity and ecosystem services. *BioScience*, 65(10), 1011–18.

Veldman, J. W., & Putz, F. E. (2011). Grass-dominated vegetation, not species-diverse natural savanna, replaces degraded tropical forests on the southern edge of the Amazon Basin. *Biological Conservation*, 144(5), 1419–29.

Vera, F. W. (2002). The dynamic European forest. *Arboricultural Journal*, 26(3), 179–211.

Vera, F. W. (2009). Large-scale nature development—The Oostvaardersplassen. *British Wildlife*, 20(5), 28.

Vera, F. W. M. (2000). *Grazing ecology and forest history*. CABI Publishing, Wallingford, Oxon, UK.

Vera, F. W. M., Bakker, E. S., and Olff, H. (2006). Large herbivores: Missing partners of western European light-demanding tree and shrub species? *Large Herbivore Ecology, Ecosystem Dynamics and Conservation*, eds Danell K, Bergström R, Duncan P, Pastor J (Cambridge Univ Press, Cambridge, UK), pp. 203–31.

Verweij, R., Verrelst, J., Loth, P. E., Heitkönig, I. M. A., & Brunsting, A. M. H. (2006). Grazing lawns contribute to the subsistence of mesoherbivores on dystrophic savannas. *Oikos*, 114(1), 108–16.

Vitousek, P. M. (2004). *Nutrient cycling and limitation: Hawai'i as a model system*. Princeton University Press, Princeton, NJ.

Vitousek, P. M., & Sanford Jr, R. L. (1986). Nutrient cycling in moist tropical forest. *Annual Review of Ecology and Systematics*, 17(1), 137–67.

Vorontsova, M. S., Besnard, G., Forest, F., Malakasi, P., Moat, J., Clayton, W. D., Ficinski, P., Savva, G. M., Nanjarisoa, O. P., Razanatsoa, J., & Randriatsara, F. O. (2016). Madagascar's grasses and grasslands: anthropogenic or natural?. *Proceedings of the Royal Society of London B: Biological Sciences*, 283(1823), p. 20152262.

Wakeling, J. L., Bond, W. J., Ghaui, M., & February, E. C. (2015). Grass competition and the savanna-grassland 'treeline': A question of root gaps? *South African Journal of Botany*, 101, 91–7.

Wakeling, J. L., Cramer, M. D., & Bond, W. J. (2012). The savanna–grassland 'treeline': why don't savanna trees occur in upland grasslands?. *Journal of Ecology*, 100(2), 381–91.

Waldram, M. S., Bond, W. J., & Stock, W. D. (2008). Ecological engineering by a mega-grazer: white rhino impacts on a South African savanna. *Ecosystems*, 11(1), 101–12.

Wallace, A. R. (1892). *Island Life; Or the Phenomena and Causes of Insular Faunas and Floras: Including a Revision and Attempted Solution of the Problem of Geological Climates*. Macmillan and Company, London.

Walter, H. (1971). *Ecology of tropical and subtropical vegetation*. Oliver and Boyd, Edinburgh, UK.

Walter, H. (1973). *Vegetation of the Earth in relation to climate and the eco-physiological conditions*. Springer, London, New York. 237 pp.

Wang, L., Jacques, F. M., Su, T., Xing, Y., Zhang, S., & Zhou, Z. (2013). The earliest fossil bamboos of China (middle Miocene, Yunnan) and their biogeographical importance. *Review of Palaeobotany and Palynology*, 197, 253–65.

Warman, L., & Moles, A. T. (2009). Alternative stable states in Australia's Wet Tropics: a theoretical framework for the field data and a field-case for the theory. *Landscape Ecology*, 24(1), 1–13.

Weaver, J. E. (1958). Summary and interpretation of underground development in natural grassland communities. *Ecological Monographs*, 28(1), 55–78.

Weigl, P. D., & Knowles, T. W. (2014). Temperate mountain grasslands: a climate–herbivore hypothesis for origins and persistence. *Biological Reviews*, 89(2), 466–76.

Wells, P. V. (1962). Vegetation in relation to geological substratum and fire in the San Luis Obispo quadrangle, California. *Ecological Monographs*, 32(1), 79–103.

Werdelin, L., & Sanders, W. J. (2010). *Cenozoic mammals of Africa*. University of California Press, Berkeley.

Werneck, F. P. (2011). The diversification of eastern South American open vegetation biomes: historical biogeography and perspectives. *Quaternary Science Reviews*, 30(13–14), 1630–48.

West, A. G., Nel, J. A., Bond, W. J., & Midgley, J. J. (2016). Experimental evidence for heat plume-induced cavitation and xylem deformation as a mechanism of rapid post-fire tree mortality. *New Phytologist*, 211(3), 828–38.

Westerling, A. L., Turner, M. G., Smithwick, E. A., Romme, W. H., & Ryan, M. G. (2011). Continued warming could transform Greater Yellowstone fire regimes by mid-21st century. *Proceedings of the National Academy of Sciences of the United States of America*, 108(32), 13165–70.

Westoby, M. (1988). Comparing Australian ecosystems to those elsewhere. *BioScience*, 38(8), 549–56.

White, T.C. R. (2005). *Why does the world stay green? Nutrition and survival of plant-eaters*. CSIRO Publishing, Melbourne, Australia.

Whitecross, M. A., Archibald, S., & Witkowski, E. T. F. (2012). Do freeze events create a demographic bottleneck for *Colophospermum mopane*? *South African Journal of Botany*, 83, 9–18.

Whittaker, R. H. (1975). *Community and ecosystems*. New York, USA: McMillan.

Whyte, I. J., van Aarde, R. J., & Pimm, S. L. (2003). Kruger's elephant population: its size and consequences for ecosystem heterogeneity. In *The Kruger experience: ecology and management of savanna heterogeneity*, Edited by Johan T. du Toit, Kevin H. Rogers, Harry C. Biggs. ISLAND PRESS Washington, Covelo, London, pp. 332–48.

Wicht, C. L. (1948). A statistically designed experiment to test the effects of burning on a sclerophyll scrub community. I. Preliminary account. *Transactions of the Royal Society of South Africa*, 31(5), pp. 479–501.

Wigley, B. J., Cramer, M. D., & Bond, W. J. (2009). Sapling survival in a frequently burnt savanna: mobilisation of carbon reserves in *Acacia karroo*. *Plant Ecology*, 203(1), 1.

Wigley, B. J., Bond, W. J., & Hoffman, M. T. (2010). Thicket expansion in a South African savanna under divergent land use: local vs. global drivers? *Global Change Biology*, 16(3), 964–76.

Wigley, B. J., Coetsee, C., Hartshorn, A. S., & Bond, W. J. (2013). What do ecologists miss by not digging deep enough? Insights and methodological guidelines for assessing soil fertility status in ecological studies. *Acta Oecologica*, 51, 17–27.

Wilcke, W., & Lilienfein, J. (2004). Element storage in native, agri-, and silvicultural ecosystems of the Brazilian savanna. II. Metals. *Plant and Soil*, 258(1), 31–41.

Williams, J. (2013). Exploring the onset of high-impact mega-fires through a forest land management prism. *Forest Ecology and Management*, 294, 4–10.

Williams, M. (2003). *Deforesting the Earth: from prehistory to global crisis*. University of Chicago Press, Chicago, USA.

Willis, K. J., & McElwain, J. C. (2014). *The evolution of plants*. 2nd edition.Oxford University Press, Oxford, UK.

Wilson, J. B., & Agnew, A. D. (1992). Positive-feedback switches in plant communities. In Advances in ecological research (Vol. 23, pp. 263–336). Academic Press.

Wilson, S. L., & Kerley, G. I. (2003). Bite diameter selection by thicket browsers: the effect of body size and plant morphology on forage intake and quality. *Forest Ecology and Management*, 181(1–2), 51–65.

Wing, S. L., & Boucher, L. D. (1998). Ecological aspects of the Cretaceous flowering plant radiation. *Annual Review of Earth and Planetary Sciences*, 26, 379–421.

Wing, S. L., Herrera, F., Jaramillo, C. A., Gómez-Navarro, C., Wilf, P., & Labandeira, C. C. (2009). Late Paleocene fossils from the Cerrejón Formation, Colombia, are the earliest record of Neotropical rainforest. *Proceedings of the National Academy of Sciences of the United States of America*, 106(44), 18627–32.

Wirth, C. (2005). Fire regime and tree diversity in boreal forests: implications for the carbon cycle. In *Forest Diversity and Function* (pp. 309–44). Springer, Berlin, Heidelberg.

Woinarski, J. C., Burbidge, A. A., & Harrison, P. L. (2015). Ongoing unraveling of a continental fauna: decline and extinction of Australian mammals since European settlement. *Proceedings of the National Academy of Sciences of the United States of America*, 112(15), 4531–40.

Wood, S. W., & Bowman, D. M. (2012). Alternative stable states and the role of fire–vegetation–soil feedbacks in the temperate wilderness of southwest Tasmania. *Landscape Ecology*, 27(1), 13–28.

Woodward, F. I. (1987). *Climate and plant distribution*. Cambridge University Press, Cambridge, UK.

Woodward, F. I., & Lomas, M. R. (2004). Vegetation dynamics–simulating responses to climatic change. *Biological Reviews*, 79(3), 643–70.

Woodward, F. I., Lomas, M. R., & Kelly, C. K. (2004). Global climate and the distribution of plant biomes. *Philosophical Transactions of the Royal Society of London. Series B: Biological Sciences*, 359(1450), 1465–76.

Wright, I. J., Reich, P. B., Westoby, M., Ackerly, D. D., Baruch, Z., Bongers, F., ... & Flexas, J. (2004). The worldwide leaf economics spectrum. *Nature*, 428(6985), 821.

Wulf, A. (2015). *The invention of nature: Alexander von Humboldt's new world*. Alfred A. Knopf, New York.

Wyse, S. V., Burns, B. R., & Wright, S. D. (2014). Distinctive vegetation communities are associated with the long-lived conifer *Agathis australis* (New Zealand kauri, Araucariaceae) in New Zealand rainforests. *Austral Ecology*, 39(4), 388–400.

Xu, C., Hantson, S., Holmgren, M., Nes, E.H., Staal, A., and Scheffer, M. (2016). Remotely sensed canopy height reveals three pantropical ecosystem states. *Ecology*, 97(9), 2518–21.

Xu, C., Vergnon, R., Cornelissen, J. H. C., Hantson, S., Holmgren, M., van Nes, E. H., & Scheffer, M. (2015). Temperate forest and open landscapes are distinct alternative states as reflected in canopy height and tree cover. *Trends in Ecology & Evolution*, 30(9), 501–2.

Yelenik, S. G., Stock, W. D., & Richardson, D. M. (2004). Ecosystem level impacts of invasive *Acacia saligna* in the South African fynbos. *Restoration Ecology*, 12(1), 44–51.

Zaloumis, N. P., & Bond, W. J. (2016). Reforestation or conservation? The attributes of old growth grasslands in South Africa. *Philosophical Transactions of the Royal Society of London B: Biological Sciences*, 371(1703), 20150310.

Zhao, K., & Jackson, R. B. (2014). Biophysical forcings of land-use changes from potential forestry activities in North America. *Ecological Monographs*, 84(2), 329–53.

Zimov, S. A., Chuprynin, V. I., Oreshko, A. P., Chapin III, F. S., Reynolds, J. F., & Chapin, M. C. (1995). Steppe-tundra transition: a herbivore-driven biome shift at the end of the Pleistocene. *American Naturalist*, 146(5), 765–94.

Index